3rd Biennial Conference of the Interdisciplinary Society for Quantitative Research in Music and Medicine (ISQRMM 2015)

Published in the Journal of the Interdisciplinary Society for Quantitative Research in Music and Medicine

Immaculata, Pennsylvania, USA
24-26 July 2015

Editor:

Joseph Ruszkowski

ISBN: 978-1-5108-3189-6

Printed from e-media with permission by:

Curran Associates, Inc.
57 Morehouse Lane
Red Hook, NY 12571

Some format issues inherent in the e-media version may also appear in this print version.

ISBN: 978-1-5108-3189-6

Printed by Curran Associates, Inc. (2016)

For permission requests, please contact ISQRMM
at the address below.

ISQRMM
c/o David Akombo, Ph.D.
Jackson State University Department of Music
P.O. Box 17055
Jackson, MS 39217

http://isqrmm.org

The *Journal of the Interdisciplinary Society for Research in Music and Medicine* is a peer-reviewed journal presenting scholarly research into music's connection and application in medicine. Articles are submitted from scholars who have presented at the society's biannual conference, as well as from the field. Inquiries regarding submission should be directed to contact@isqrmm.com

Additional copies of this publication are available from:

Curran Associates, Inc.
57 Morehouse Lane
Red Hook, NY 12571 USA
Phone: 845-758-0400
Fax: 845-758-2634
Email: curran@proceedings.com
Web: www.proceedings.com

Interdisciplinary Society for Quantitative Research in Music and Medicine

Proceedings Journal 2015

President's Forward

The study of the beneficial effects of music in medicine is a discipline that has spurred an interest among scholars from both the humanities and biomedical disciplines since the 18th century. Researchers in the 20th century embraced these studies with scholarly and academic vigor. Those in the 21st century have continued on with this inquiry. The mission of the Proceedings *Journal of the Interdisciplinary Society for Quantitative Research in Music and Medicine* is to provide global readership of peer reviewed articles that explore the unanswered questions regarding the efficacies of all forms of music on the human body, mind and soul. The Journal disseminates ISQRMM's biennial conference proceedings as well as new, unpublished and independent research papers not published or presented at the conferences. The Journal welcomes qualitative research as well as quantitative research methodologies and encourages interdisciplinary formats. The Journal encourages the authors to utilize modes of research that are clearly defined, explained and relevant. The Journal also welcomes Master's or Doctoral theses in any related fields.

The topics of historical qualitative research, regarding music and healing, especially as revealed in its manifold and remarkable manifestations—throughout Ancient Greece until the dawn of the 18th century—has never been candidly and fairly examined. There is also a dearth of research spanning the global cultures which are worthy of examination. Presently, the only sound sources of information on the culturally diverse aspects of music as medicine are anecdotes, which provide eyewitness accounts of individual music healers. In addition to the quantitative studies in the field, the anecdotal repository of evidence, as recorded in treatises of healing rituals, add credence to the historical implications of music as a healing agent.

The *Journal of the Interdisciplinary Society for Quantitative Research in Music and Medicine* balances its readership and uses an eclectic model to select the research papers submitted to make the Journal both scholarly and informative for all readers regardless of specialization. We are proud of ISQRMM supporters, enthusiasts, and our global clientele of our dedicated team of scholars.

David O. Akombo, Ph.D.

Co-Founder and Board Member.

Editor's Forward

This proceedings journal represents an array of articles not only from the field of Music Therapy and Musicology linked by a common thread of quantitative analysis, but from a broad swath of academic disciplines. In this age of increasing specialization in academia, there is a need for bridges to be built across disciplines, thereby allowing fresh interpretive models to be applied to the data expanding the intellectual landscape of the understanding of music. The Interdisciplinary Society for Quantitative Research in Music and Medicine has provided a platform for an array of scholars, clinicians, and practitioners to assemble and explore the healing aspects of music. This journal gathers articles reflecting the breadth of approach that comprises the research of the society. As the editor, it has been my pleasure to review the articles for inclusion in the journal. I am impressed by the level of scholarship presented herein, and I am certain that the reader will be challenged to expand his horizons and view the field in new and inspiring ways. It is my earnest hope that this proceedings journal will lead to further communication and discussion with the authors, and the further development of continued research at the society's next conference.

Steve Jackowicz, Ph.D.
Board Member

Contents

Mandala as Recovery Tool

Mary Elizabeth Campbell, MAMT-BC

Abstract

I developed my Mandala in Recovery protocol for use in combination with music therapy groups for Intensive Outpatient (IOP) mental health consumers, and further developed it in combination with EMDR as a tool for clients in trauma recovery. The creation of a personal Recovery Mandala by each client was the culmination of a series of IOP recovery groups designed to promote 'High Level Wellness' and present wellness as holistic, integrating all sides of the health triangle of physical-medical/mental-emotional/spiritual-social. The Recovery Mandalas are very tangible, creative and highly individual expressions of core positive beliefs, goals, hopes and dreams, and are decorated with colors and design choices of the client. They serve as visual concrete reminders of progress, goals, dreams and ideals. They may include many things for which the person is thankful, and can be a powerful tool in self-care and ongoing recovery. In the context of EMDR the creation of a Recovery Mandala was a very natural outgrowth of a series of sessions in which an integral part of the process is to move towards positive beliefs about self while trying to achieve a holistic recovery from trauma. Clients in both groups report enjoying having the Mandala available as visual inspiration, emotional support, and as personal art work. Many IOP clients chose to leave their Mandalas on display in the facility as continued support and inspiration, and to also inspire peers. All clients are encouraged to use the Mandala as a meditation focus along with preferred music for relaxation.

Introduction

My use of the Mandala form grew from my appreciation of it in undergraduate school where I majored in Religion Studies and Music Performance, and minored in Psychology and Fine Arts. At that point I saw the Mandala as a fascinating meeting point for all those pursuits. This is not a scholarly article on the history of 'Mandala,' nor is it influenced by other practitioners who have incorporated the form in their work, but I will provide some basic background.

'Mandala' is a term essentially meaning sacred circle pattern which I first remember encountering while studying some of the more mystical aspects of Eastern religious traditions. 19th century English explorer and researcher L. Austine Waddell, renowned for his extremely comprehensive writing on Tibetan Buddhism, called them 'Magic Circles.'[1] Beautiful hanging cloth paintings, or Thanghkas, were created that represented the universe in Mandala form, and contained very intricate detail – these were then used as a focal point for meditation. The term Mandala is Sanskrit meaning circle and originally was associated with Hindu and Buddhist graphics symbolizing the universe, generally including four quadrants. There are many examples in Hindu and Buddhist iconography, and they are stunning in their intricacy and beauty. Mandalas now are by no means limited to those traditions and quite often they can be associated with a mystical reaching for the divine, for peace, for oneness, for understanding the universe. One practice by Tibetan Buddhist monks is that of creating an incredibly beautiful and complex sand Mandala, full of many levels of symbolism, intricacy and detail, and then destroying the Mandala a short time after completion as a symbol of impermanence.

The term Mandala is often now generalized to mean circular geometric patterns, and there is still a sense of representation of the universe, the mystical, and the eternal. The great psychologist Carl Jung was fascinated by the Mandala and Peter Patrick Barreda, in his on line essay 'Archetype of Wholeness: Jung and the Mandala,' writes

"... Jung refers to the Mandala as 'the psychological expression of the totality of the self.' ...the Mandala is the template for the mind, a state of peace and order, resolution of the chaos within..."

In his autobiographic work Memories, Dreams and Reflections – 1963 Carl Jung describes his journey through PTSD after witnessing the atrocities of World War I. A major part of his recovery involved taking time each day to sketch and journal, and he describes drawing Mandalas as key in his process of returning wellness. Robert M. Johnson, PhD, describes Jung's journey in his book Life's Full Circle: The Mental Health Self-Help Series. He quotes Jung saying, "My Mandalas were cryptograms...in which I saw the self – my whole being...." Jung went on to study patients with major mental illness and found that there was a universality to drawing of the circle. He felt that when a person is feeling shattered, the Mandala can help them regain wholeness and rebuild. Various people have discovered the Mandala within the context of therapy, including those who suggest specific interpretations of color and shape placement vis-à-vis therapeutic process.

I would suggest that the idea of sacred circle, while perhaps not always as intricate as those done in Eastern iconography, appears worldwide and has a fundamental grounding appeal. In Christianity we have the Rose windows, beautiful circular stained glass windows.

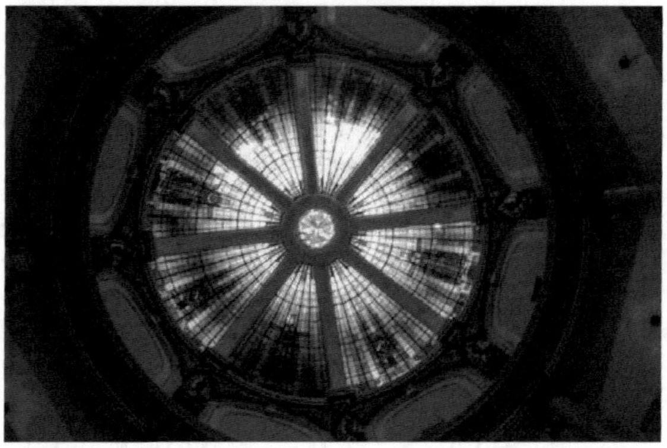

The window pictured here is in the dome over the main hall at Immaculata University and is a beautiful example of a stained glass Mandala of Christian iconography. There is also the tradition of the Labyrinth, which involves physically walking within patterns in a circle until reaching the center, and which is done as prayer and as meditation, sometimes in lieu of a long religious pilgrimage. The labyrinth pictured below was created in Phoenixville in 2005 as part of a community garden with work done by adjudicated youth – I was honored to be able to provide music for its opening and involved many of those youth. When it was complete, herbs were planted along the edges and a stone bench sat at the center. Members of the surrounding community were encouraged to use it for peaceful contemplation.

There are also the Amish Hex signs, Celtic Knots, circular patterns within Native American iconography, and so many more examples. I would also submit the idea of the Hora and other traditional dances that are circle based – yet another example, but in motion, of the powerful circle pattern. And continuing with circles in motion, the Sufi sect of the Whirling Dervishes which began in the 12th century is very much part of a mystic tradition trying to reach towards the divine, the eternal – all while moving in circles!

Method

Inspired by the concept of Mandala, I developed my own process for using them in recovery. Having worked with survivors of abuse and trauma, as well as with those suffering with major mental health diagnoses, I've seen many people who have enormous self doubt and shattered self confidence. I have observed that these individuals tend to have chronically poor self image and they question themselves at every turn, and I wanted to find a concrete, tangible way to help begin to rebuild a sense of self worth, while reinforcing work done in therapy. The series of IOP group sessions leading to the creation of Recovery Mandalas included the following:

- Each group sought to help clients build recovery tools and often included music therapy as an integral part of the process.
- Specific examples of group sessions included identifying and using music with movement to increase exercise both during group and at home; group process choreography for

community building, movement, stress release; exploring the impact of music on emotion and encouraging conscious use of musical choice for supporting emotional wellness; group music improvisation for exploring emotions, self expression, stress relief and awareness of others; song writing and song analysis to help further explore various recovery issues often with a touch of humor; guided relaxation with music and building skills for self soothing and stress reduction.

- Prior to the session in which clients began creating Recovery Mandalas, I combined guided meditation and positive visualization with music chosen to promote relaxation, and clients were encouraged to imagine being in a safe, comfortable, affirming place, and to then think what they most wanted to believe about themselves. This was followed by group discussion of possible affirmations and goals, and the group generated a long list of affirmations and positive directions, very much inspired by one another.

On Generating Affirmations:

Most of us are very good at self criticism and internalizing the negative, and especially so for survivors of abuse and various forms of trauma and chronic mental health issues. After working to achieve building blocks towards recovery, a very helpful step can be finding core positive beliefs, or affirmations. Participants are asked: "What core belief DO you, *or would you LIKE to,* believe about yourself?" Then they are guided through the following:

- Participants are led in guided meditation with music during designed to elicit relaxation, and then are invited to imagine a safe, healing, and accepting place, and then are invited to think of affirmations. In a group setting I tend to use David and Steve Gordon's Shaman's Vision Journey, Steven Halpern's Inner Peace, or similar music intended to help achieve relaxation. This is partly to avoid lyric distractions and possible triggers, and to use something that is as close to universally relaxing as possible.

- If it's too hard to name positives for yourself, what might you say to a loved one? A friend? Perhaps imagine what you as your adult self might say to your child self?
- Can you name cherished hopes and dreams?
- Can you name things that you are thankful for – from the tiniest appreciation to the biggest thing?
- Look at what you wrote and begin to prioritize – what are the MOST important things?
- Organize them by priority, deciding which things must be part of your Mandala

Some sample affirmations:

- **I am a survivor**
- **I'm a positive person**
- **I can be a good leader**
- **I'm loveable**
- **I'm resilient**
- **I'm competent**
- **I'm a giving and caring person**
- **I'm a kind person**
- **I am beautiful in many ways**
- **I am a volunteer**

With EMDR clients, session notes provide a record of specific positive cognitions expressed and arrived at by clients. After several sessions I may offer the option of Recovery Mandala creation if I see that real progress has been made, and if I feel that it could be a helpful tool. We can then revisit affirmations, hopes, dreams and goals as they make choices for what to include in their work. I routinely include music specifically to aid in relaxation as a part of EMDR work, carefully tailored to the needs of the individual. Within the context of IOP sessions, as clients begin to share possible affirmations they inspire one another and the list grows as a part of group process.

The specific process of creating individual recovery Mandalas includes the following:

1. Clients choose from a range of possible circular designs, all with

an open circle in the center for one's most important affirmation. I wanted clients to see the center as a point on which to orient, to ground oneself. The choice of phrase for this point is very personal and very important. Below are just two of several design choices:

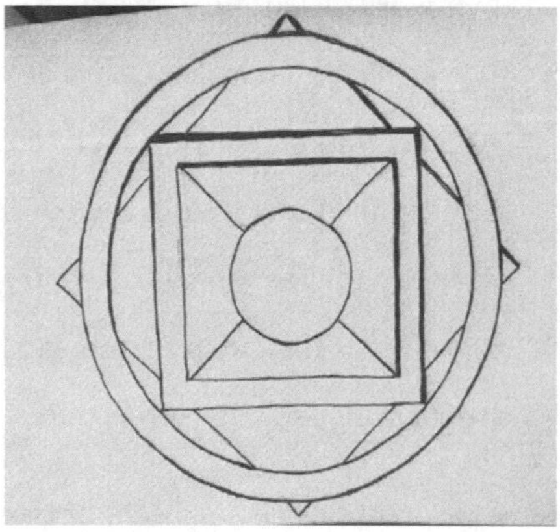

2. Clients are encouraged to choose colors and a medium (colored pencil, marker, water color, crayon...) that they resonate best with, and to begin to add design and color elements to the Mandala.
3. Clients write their most important affirmation in the center circle and then incorporate other affirmations, goals, hopes and dreams within the body of the Mandala. Sometimes clients may use the words themselves as the design process, doing text in various colors and styles.
4. Clients are encouraged to be as creative as possible, adding drawing or geometric elements as they wish, printing the text in whatever way they choose, with facilitator assisting and supporting if need be.
5. When the Mandala is completed to the client's satisfaction, they are given the option to frame. This involves choosing a fabric which then covers the cardboard frame.
6. Clients are encouraged to use the Mandala for personal

meditation coupled with music for relaxation.

The process of creating a Mandala can take several sessions, and I recommend using music for relaxation during their creation. If working with an individual I would present them with several choices of music that could be played, and I would have worked with that person to determine the types of music that they might find most relaxing and/or supportive. Sometimes a client may be very unsure of their ability. The geometric grids can provide a soothing and relatively easy format for simply adding color if a person feels that they are limited artistically, and the end results can be quite pleasing regardless of ability. For those who are artistically gifted, the works can become quite intricate and lovely. Occasionally I have assisted people who could not physically do the printing or coloring, but who took an active role in choosing colors and all design placement ideas.

Results

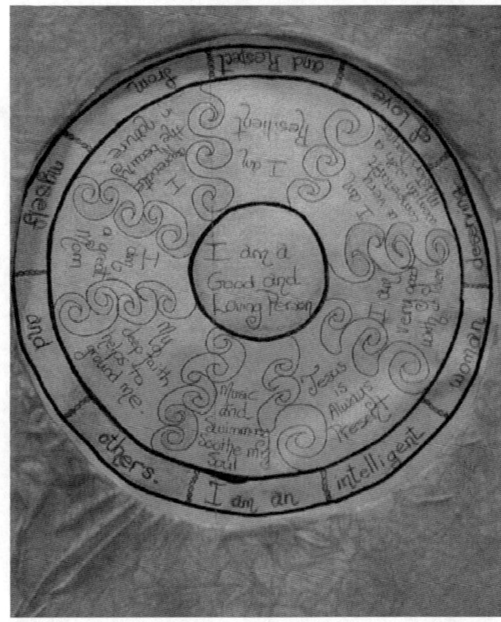

This Mandala was developed with and for an EMDR client who suffered both repeated abuse as a child and domestic violence. As is often the case with abuse victims, she struggles with constant self doubt. She didn't trust herself to do the Mandala alone, but did work closely with all the details of it, including the exact color of the frame. As soon as it was complete she decided it looked perfect and that she wanted to work on another one by herself. She uses the Mandala as a meditative focus as she works with basic meditation and self soothing techniques. I prepared a CD for her with relaxing music and a voice over of some guidance in relaxation.

IOP clients very enthusiastically took part in creating their Recovery Mandalas. Several were quite artistic and found them to be a wonderful outlet for their creativity, but even those who were limited in that regard found the geometric patterns to be easy to work with and be successful with. There was a real sense of pride with the creation of the works, and participants were very engaged in also choosing fabric and frame designs to best display and protect their work. The Mandala project was so well received in the agency that many were put on display, with client permission, as part of Mental Health Awareness Week, and they remain on agency walls. The following are examples of client Mandalas:

This Mandala was created by an IOP client who consistently worked hard in group and found this exercise to be very affirming and supportive of her treatment goals in general. I'm pleased to report that she has subsequently gotten a part time job and moved into her own community apartment.

This IOP client chose to have the words themselves create the colorful pattern for his Mandala. The frame combines fabric, card board and duct tape. This man was very quiet in groups and tended to keep to himself, but got wonderful feedback from others about his creation, and seemed genuinely pleased.

This IOP client, also an abuse survivor, brought her considerable artistic abilities to the task at hand and created a stunning and motivating piece. Art is very important in her recovery process and this was a welcome part of that. She even added levels of creativity to the frame creation, choosing to use two fabrics plus duct tape together.

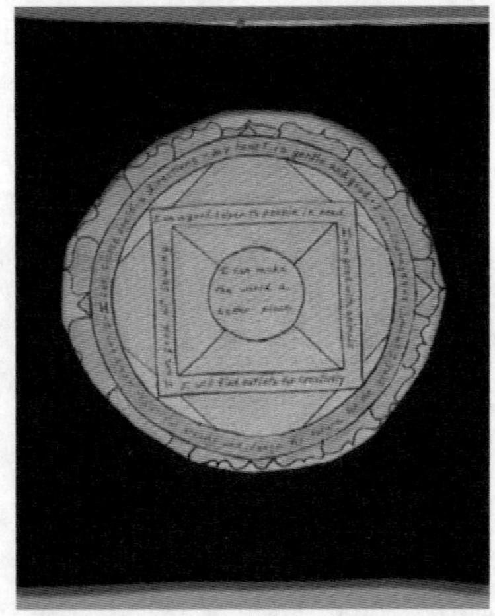

This IOP client chose marker and colored pencil, with a nice combination of colors and writing within a simple pattern. Her choice of frame fabric really sets the image off. This is one of many now hanging on the agency walls with client permission.

This IOP individual struggles with literacy and was not confident in his ability to write, so I offered the option of lettering that he could add color to. His design choices are simple and strong, and he was very proud of the finished product, which he showed to peers and therapists alike.

This IOP client had an interesting design process - instead of covering our cardboard pizza box with fabric for the frame, she did a design that reflected the box itself. She chose to use images in the center space and have her affirmations in the next concentric circle. Several people did this - some chose a butterfly or birds to represent recovery on many levels.

This Mandala was done by a special needs older man in a hospice setting – he dictated exactly how he wanted it to appear, and exactly what he wanted it to say. After his death I had occasion to use it at his funeral as a jumping off point for words of remembrance, and a way for him to speak to his friends one last time, which wound up being a more profound use than I ever anticipated.

I presented this protocol to 'Women for Sobriety' support groups as a single 2 hour group session. This included a list of possible affirmations and a discussion of finding a core affirmation or positive truth for your own recovery, and then hands on Mandala creation with a variety of possible medium choices (crayon, colored pencil, marker, water color.) Participants reported that the list of affirmations and related discussion fit very well within their 12 Step model, and that the physical act of creating the Mandala was in itself very relaxing.

Discussion

I have found the Recovery Mandala to be a very helpful tool with a variety of clients, from chronic mental health consumers, to people recovering from severe trauma. Mandalas can be a comfort, a source of inspiration, a summary of hope and progress. I also found the Mandala to be a very positive application as goal support and a positive intervention with Developmentally Disabled clients – in one instance I happened to make one with a DD man in hospice and it hung on his hospital wall to help introduce him to his care team, and I used it to speak for him during his funeral; I also made one with a dually diagnosed DD/MH man and it has hung prominently in his group home ever since. The design of the core positive belief surrounded by goals, beliefs, things one is thankful for, may well resonate with a range of populations. I feel that my clients have found them to be very helpful, becoming a tangible reminder of goals, progress, hopes and dreams. It may be best summarized by one of my IOP clients, who said, "I have my Mandala on my wall at home. I look at it when I'm upset and it calms me down. It helps remind me of how far I've come."

References

Barreda, Peter Patrick. (2009) Archetype of Wholeness: Jung and the Mandala *from Essays on Mandalas, Spirituality and the Universe,* retrieved May 17, 2015 from http://www.wakingtimes.com/2015/05/15/archetype-of-wholeness-jung-and-the-mandala/

Fincher, Susanne F. (1991, reprint 2010) *Creating Mandalas*. Boston, Mass: Shambala Publications, Inc..

Gordon, David and Steve. (2006) *Shaman's Vision Journey*, Topanga, CA: Sequoia Records, Earth Vision Music (BMI).

Halpern, Steven. (1994) *Inner Peace*, San Anselmo, CA: Sound Rx, Open Channel Sound Co.

Johnson, Robert M., PhD. (2014). *Life's Full Circle: The Mental Health Self-Help Series.* Palmyra, VA:Creative Publications.

Waddell, L. Austine (1972). *Tibetan Buddhism With Its Mystic Cults, Symbolism and Mythology*. New York, New York: Dover Publications, Inc.. Republication of 1895,

The Buddhism of Tibet or Lamaism. London, England: W. H. Allen & Co.

Spiritual Healing and Exorcism: Music as a Catalyst in Sacred Space

D. Weldon Cochren, M. Div.

Abstract

The focus of this paper is to explore the possibility of music and its causality in spiritual healing, and exorcism in the Pentecostal Church within the African Diaspora. The term "spiritual healing" as opposed to just "healing" is used to indicate 1) an event that would occur in the realm of the miraculous or super-natural; and 2) an event that should not be misconstrued as the result or process of Music Therapy, as some have perceived or concluded David's musical encounter with King Saul to have been (1 Samuel 16:23). Biblically, there is no evidence of music being incorporated into the ministry of spiritual healing or exorcism of Jesus or the Apostles, yet music has become an integral component in Pentecostal churches throughout the African Diaspora (Continental Africa, North America, Brazil, United Kingdom, Caribbean and Virgin Islands). Music for spiritual healing also is also prevalent in many communities that are not contained or confined in churches. This is evident in the countries of Haiti, Mali, and USA. This paper will endeavor to answer the following questions: Does a particular genre or style of music have a greater impact in the sacred space than others? Is there a particular spiritual position or comprehension that musicians and singers should occupy or obtain to achieve the desired outcomes? Does the cultural or worship community have an influence on the musical expression, e.g., instruments, lyrical content, language? Does meter, tempo, key, or pitch affect the outcomes? Does the "patient" have to be a Christian? Is there a preparation and treatment for the instruments used in the rituals or ceremonies? What is the training or conditioning process for the musician(s)? Conclusion: The transporting of sicknesses associated with spiritual oppression, spiritual possession, physical ailments, psychological/mental distress, or biological toxins from a human can be facilitated through or assisted by music, with certain exceptions and stipulations.

The focus of this paper is to explore the possibility of music and its causality in spiritual healing, and exorcism in the Pentecostal Church within the African Diaspora. The term "spiritual healing" as opposed to just "healing" is used to indicate; an event that 1) would occur in the realm of the miraculous or super-natural; and 2) not be misconstrued as the result or process of Music Therapy, as some have perceived or concluded David's musical encounter with King Saul to be (1 Samuel 16:23).

> 23 And it came to pass, when the *evil* spirit from God was upon Saul, that David took the harp, and played with his hand: so Saul was refreshed, and was well, and the evil spirit departed from him. [2]

Biblically, there is no evidence of music being incorporated into the ministry of spiritual healing or exorcism of Jesus or the Apostles, yet it has become an integral component in Pentecostal churches throughout the African Diaspora (Continental Africa, North America, Brazil, United Kingdom, Caribbean and Virgin Islands). It is important that a general understanding of particular portions of spirituality in the African context and how that spirituality is intertwined in the lives of people of African descent throughout the African Diaspora. To a degree music, can be used as a segment of a "deliverance" process. In the book of Acts (16:25), when Paul and Silas were in prison and sang hymns, they caused an earthquake to occur, wherein the doors of the prison opened.

> 25 But about midnight Paul and Silas were praying and singing hymns unto God, and the prisoners were listening to them; 26 and suddenly there was a great earthquake, so that the foundations of the prison-house were shaken: and immediately all the doors were opened, and every one's bands were loosed. [3]

Without the singing, we can only speculate on what the outcome would have been. In most cases, worship will precede a visitation of the Spirit of God or the manifestation of miracles. Having said that, if one were to research aspects of African-based religions and Pentecostalism, one would understand that ritual will precede worship or ritual is encased in worship: shouting, dancing, doing call and response,

---

[2] *American Standard Version*. Oak Harbor, WA : Logos Research Systems, Inc., 1995, S. 1 Sa 16:23.

[3] *American Standard Version*. Oak Harbor, WA: Logos Research Systems, Inc., 1995, S. Ac 16:25.

bearing witness, hand clapping and foot stomping. The premise of this paper is that music is a catalyst in sacred space for healing and exorcism. The emphasis on "sacred space" in this context is relegated to a location dedicated to Christian worship because others practice ritual in a non-Christian "sacred space." There is an old proverb, "There are no atheists in Africa." Spirituality is central in the lives of the people no matter where you visit. In many countries on the continent of Africa, there is a strong belief in communing with ancestral spirits and various other spiritual belief systems, e.g., Yoruba (Nigeria), Vodou (Haiti), Santeria (Puerto Rico), Candomble (Brazil), and Rara (Louisiana, (USA)). There is a common musical instrument that is central in the aforementioned spiritual belief systems and in much of the culture of African peoples, be it religion, communications, warfare or celebrations. It is the drum.

Dr. David Akombo writes in his book, *Music and Healing Across Cultures*, "Africans generally tend to approach healing through music and dance."[4] Yaya Diallo and Mitchell Hall, in their text, *The Healing Drum: African Wisdom Teachings,* state, "In the Minianka villages of Fienso and Zangasso, the musicians were healers, the healers musicians. The word musician itself implies the role of healer. From the Minianka perspective, it is inconceivable that the responsibilities for making music and restoring health should be separate, as they are in the West."[5]

The Drum, or more accurately, drums are used in ritual for the purpose of summoning spirits. The drums are usually arraigned in configurations of 3, not by accident, rather by design. In some religions like Vodou, the drums are arraigned with the Trinity in mind, although the worship service has no direct connection to Christianity. Often, the drums have been dedicated to the service of invoking spirits through a ceremony which includes the pouring of libations, mostly peppermint schnapps, sometimes water. It is also strongly believed that the drums have spirits. "Because Vodou drums are sacred, ritual activities are necessary for their creation and maintenance. The

[4] *Music and Healing Across Cultures*, 79.

[5] *The Healing Drum: African Wisdom Teachings*, 79.

proper care of drums enhances their power to call the spirits."[6]

"Among today's Vodou servants in Port-au-Prince and New York, Outo or Outogi signify the presiding spirit of the drums."[7] Furthermore, "A special instrument for women is the water drum, which is made from a vessel filled water and covered with an inverted calabash."[8]

The drummers are extensively trained in rhythms that are used to invoke spirits. The drummer is a primary conduit for spirit possession, being careful that he (female drummers are rare) does not become possessed in the process. Spirits are perceived to have the ability to impart sicknesses on individuals. It is important to note that spirit possession in this context involves a spirit "mounting" a human as opposed to Holy Ghost possession where an "indwelling" of a human occurs. Music is a spirit; it cannot be seen or touched. Yet, music can be felt, is a transporter of emotions and creates a desired environment.

The music in sacred space is not necessarily selected by genre nor is it genre driven. Many in Christendom, both church and academy, seem to have a need to categorize music into nice, neat little compartments, then proceed to asses each compartment based on dogma and personal taste, which is usually dictated by one's religious doctrine. Certain songs by Bob Marley, e.g., "Redemption Song," or "One Love" or the prophetic voices of Curtis Mayfield, "People Get Ready," or "New World Order" can be ignored because they are not, "one of us."

It is not so much a selected genre or particular style of music, as it pertains to instrumental verses vocal, tempo or rhythm, key or mode that facilitates spiritual healing. Although to some degree, those factors may be important when we consider healing and exorcism and not just worship intensity. "Spirit possession involves (1) drums and (2) dance... (16) Music varies in tempo and complexity of polyrhythms in correspondence with various spirits."[9] The primary consideration

[6] *Vodou Drums*, 44.
[7] *Vodou Drums*, 44.
[8] *The Healing Drum: African Wisdom Teachings*, 81-82.
[9] "Can't Nobody Do Me Like Jesus", 84.

is that the songs are Spirit composed and Spirit led. In other words, the songs should be Spiritual songs.

> 18And be not drunken with wine, wherein is riot, but be filled with the Spirit; 19speaking one to another in psalms and hymns and spiritual songs, singing and making melody with your heart to the Lord;[10]

The primary goal is to invoke the Holy Spirit and for the Holy Spirit to move on the person's behalf (healing and/or exorcism) through the song. Music, by and large, is not created in a vacuum. An individual or group of musicians or singers, or both, must be aligned with the Holy Spirit and be in "spiritual concert" with one another. Being in "concert" musically is secondary to being "spiritually cohesive." The musician can be formally or informally trained, read music or play "by ear," have theoretically correct voicing or the presence of enharmonic in a horn or string section is not a major concern, but admittedly a sour note is a sour note. Everyone's instrument should be tuned properly and play in the same key or the proper key transposition. The Spirit can manifest a song through a musician or singer, comparatively speaking, as an interpreter receives and manifests a communication from the Holy Spirit having heard glossolalia.

The drum is an important instrument in both African and Pentecostal ritual. The drums significance is such that during the slave era on the North American mainland and throughout the Caribbean, it was outlawed. In the African Diaspora drums are central in Pentecostal and non-Christian ritual. The drum has also found its purpose in the various offspring of African religious based and Pentecostal music, e.g., R&B, Rock & Roll, Blues, Jazz, Funk, Soca, Dancehall, Ska and Hip Hop.

Just as the drum(s) and African based music have extracted a particular response or reaction from the general society, the usage or non-usage of certain instruments, including the drum is largely based on the culture of the particular worship community and its socio-political, theological position. For example, if a worship community prides itself on residing in and being frequented by people of an

---

[10] *American Standard Version*. Oak Harbor, WA : Logos Research Systems, Inc., 1995, S. Eph 5:18.

upper income status, and has staunch European-American "mainstream" theology, regardless of race or ethnicity, one would be hard pressed to find musical instruments other than organ, piano and the singing of hymns or an occasional anthem. Those worship communities usually are not open to Pentecostal based pneumatology. Many worship communities are ridged in their ritual of liturgy, which prohibits the infusion of Spiritual songs or a move of the Holy Spirit. Unfortunately, their pneumatology is based on the "warm and fuzzy" perspective, void of any tangible, demonstrable power, "kindom of God," as opposed to a "kingdom" or "Empire of God." Any worship community that views the music and song that is utilized in sacred space, during ritual and worship as an element of "entertainment," consciously or subconsciously, may never comprehend nor fully appreciate the power they holds and their potential for deliverance and freedom.

Spiritual song lyrics are not required to rhyme and usually do not. Often, they have repeated the same phrase multiple times. The music is composed in the Spirit, meaning that a person receives a song by way of divine influence or impression. The song appears in the natural realm as improvised because the listener may not be aware of what occurred spiritually. In essence, a worship community should attempt to be musically open to a musical move of the Holy Spirit. That is not to say that the Spirit will bring a Heavy Metal based song to a worship community of Ghanaian immigrants.

If we understand glossolalia as Spirit to spirit communication primarily and the infusion of interpretation through a human oracle for the benefit of human understanding, then it is plausible that Spiritual songs, which are Spirit, are also engaged in Spirit to spirit communication wherein the Spirit of God can communicate through musical instruments to a human spirit with lyrics being the interpretation. The Spirit through the music maybe saying "be healed" or "come out" without lyrics, yet the miracle occurs. In her published dissertation, Dr. Thomasina Neely in an interview with Elbernita (Twinkie) Clark-Terrell and Rev. Bobby Henderson she writes:

> "Yes you can prophesy in music without having to say a word. It's an unknown language that God sends

upon you and the angels by which when you play instruments you're talking, even without words and you usher in the word of prophecy in the service or in the setting". (Clark-Terrell 1991)[11]

Concerning King David,

"David taught the people when you praise God, clap. When you praise God, shout. When you praise God lift your hands. He taught them when you praise God play the organ. Play on the high sounding cymbals. He taught them to play musical instruments, but he said when you do it, prophesy. How do you prophecy on musical instruments when musical instruments can't talk English? But they speak forth. One aspect of prophecy is not just to foretell but to speak forth and the anointing will rest on the musician and the musician can speak forth. That's how David was able to play the harp and the music told the Devil, turn him loose and the demon had to come out." (Henderson 1988) [12]

When one operates from the position, that God has created everything that in both visible and invisible to the eye it is understood that God is able to minister to all. If a person has enough belief to enter sacred space, healing and exorcism can occur. If someone is willing to enter sacred space on behalf of someone else, healing and exorcism can take place. For example, in the Book of Acts, it reads that God worked special miracles through Paul wherein aprons and handkerchiefs that were on his person contained healing power for distribution to those who were sick and possessed,

[11]And God wrought special miracles by the hands of Paul: [12]insomuch that unto the sick were carried away from his body handkerchiefs or aprons, and the evil spirits went out. [13]

Consider the possibility that the same is possible with music in the sacred space. The possibility of placing articles of clothing that may or may not belong to the "patient," into the sacred space, in the midst of music, ritual, worship and Spiritual song and may extract similar or greater results. As previously mentioned, drums are used in African based religion libations or a dedication service. Are people anointing their drums and the other musical instruments and voices that occupy sacred space?

[11] BR&P, 209.
[12] BR&P, 209.

[13] *American Standard Version*. Oak Harbor, WA : Logos Research Systems, Inc., 1995, S. Ac 19:10.

Spiritual music/songs usually follow the "norm" of music theory that is common to that particular culture. If the event occurred in an African or Asian setting, meter may not be recognizable when analyzed through the lens of Western Music Theory. There are distinctions in tuning of instruments, choir voicing, scales and modes. In Elizabeth McAlister's book, *Rara Vodou, Power, and Performance in Haiti and Its Diaspora*, she discusses the findings of ethnomusicologist Gage Averill, who according to McAlister was the only ethnomusicologist (at the time of publishing) that has analyzed Rara music. In the area of drum tuning, McAlister writes, "He [Averill] notes that while the *vaksin* are tuned in minor thirds; the two highest-pitched *vaksin* (and smallest in length) will be tuned a minor second apart. The melody that they play will often form arpeggiated diminished triads with a minor or diminished seventh."[14] In the area of vocal parts McAlister continues, "Averill found them to be pentatonic or hexatonic vocal melody. The pentatonic Rara melodies "often studiously avoid making much reference to the *vaksin* tonality." This means that two tonalities are present in the same music form."[15]

Lois Wilcken's *The Drums of Vodou*, lists the findings of Claude Dauphin's music analysis of music in Vodou and her [Wilcken's] analysis of the same. First Dauphin (translated for French), details five scale types:

> (1) Anhemitonic pentatonic; (2) diatonic heptatonic (major, natural minor, and some Dorian and Phrygian modes created by pitch alteration); (3) incomplete scales (truncated heptatonic, i.e., pentachords and hexachords); (4) tetratonic; and (5) metabolic (in two modes alternately, or at the same time).

Wilcken states, "My own collection of songs reveals similar categories. A striking number of the use this pentatonic scale: c d e g a The "tonic" may be c and/or a. Notice that there are no minor seconds in this scale. Some songs add a sixth pitch, b, but it usually has the character of a passing note and is part of a descending motif more often than an ascending one."[16]

Diallo and Hall write, "Psychological illness means being out of harmony, and

[14] *Rara*, 47.
[15] *Rara*, 47.
[16] *Drums of Vodou*, 95.

healing is the means of bringing the individual back into harmonious relation with the world...The music is polyrhythmic, which means that different rhythms are played simultaneously, with their accentuated patterns often crossing one another rather than being synchronous...This rhythmic complexity is blended with corresponding melodic and harmonic complexity. Minianka scales are pentatonic, and repeated melodic phrases are subtly varied."[17]

Dr. Thomasina Nealy gives her analysis of traditional praise songs in the Church of God in Christ (COGIC) and in an interview with the daughter of Bishop Charles H. Mason founder of COGIC, Lelia Mason-Byas analysis of song in the early days of COGIC. Nealy writes, "In traditional COGIC praise songs, there is a minor tonality with scale patterns that evolve out of the blues-based prayer moan superimposed over the major diatonic scale. On the other hand, *charismatic praise and worship* songs most frequently appear in the major diatonic scale that is not blues bases."[18]

Mason-Byas states, "In the early days, the Spirit gave you songs. Songs just came forth as the Spirit gave you words and all. Today, songs are composed. It makes a big difference because if you get one idea you got to search for rhyming words to end it with and it's nothing of the Spirit at all. You see that wasn't composed, the Lord brought that forth and it makes sense. It has reasoning. It has meaning there." (Mason Byas 1998)[19]

Clearly there are stark contrasts, yet there is room for continued investigation and even experimentation. There is a possibility that elements found in Rara or Vodou can be manipulated, countered, or altered to possibly reverse or undo possessions, recall or repel spirits related to sickness. If people are willing to admit it, the church, the believer, the scholar are engaged in spiritual warfare and that it is their responsibility to arm the body with more weapons for the conflict. Increasingly there are sicknesses, disease and conditions which have no relief or cure. The ailments are physical, physiological, neurological, social, and environmental. Many conditions

[17] *The Healing Drum: African Wisdom Teachings*, 80.

[18] BR&P, 172.

[19] BR&P, 175.

have spiritual components that can be corrected or eradicated. For many, it is difficult to understand music outside of an entertainment or commercial construct or context. If a particular music causes spirit possession and, along with that possession, sickness, the opposite is also plausible, that there is a music which can dispossess and heal. It is therefore incumbent upon participants to make available the difference in sacred space.

References

American Standard Version. Oak Harbor, WA: Logos Research Systems, Inc., 1995, S. 1 Sa 16:23

McAlister, Elizabeth, *Rara!: Vodou, Power, and Performance in Haiti and Its Diaspora.* Berkeley: University of California Press, 2002.

Neely, Thomasina, "Belief, Ritual, and Performance in a Black Pentecostal Church: The Musical Heritage of the Church of God in Christ." PhD diss., Indiana University, 1993.

Scandrett-Leatherman, Craig. "Can't Nobody Do Me Like Jesus": The Politics of Embodied Anesthetics in Afro-Pentecostal Rituals. PhD diss. University of Kansas, 2005.

Wilcken, Lois featuring Frisner Augustin, *The Drums of Vodou.* Tempe: White Cliffs Media Company, 1992.

Akombo, David Otieno, "*Music and Healing Across Cultures*". Culicidae Press, 2006.

Diallo, Yaya and Hall Mitchell, "*The Healing Drum; African Wisdom Teaching*". Vermont: Destiny Books, 1989.

The Effects of Music on the Quality of Life of Healthy Elderly People

Naejeong Suh & Roy Kennedy

The University of Georgia

Abstract

The purpose of this study was to identify optimal strategies of music intervention for contributing to the mental well-being of healthy elderly people. An experimental study was conducted to evaluate the effect of music therapy and music education on three groups of participants: healthy elderly unaccompanied during treatment sessions, healthy elderly accompanied by grandchildren (ages five to seven) during treatment sessions, and a control group. ANOVA and Independent Sample T-test techniques were employed to compare the effects of music activity among the groups, while Paired Sample T-test techniques were employed to test the effects of music activity within groups. The results indicated that the unaccompanied elderly group showed significant improvements on mood, self-esteem, and depression measurements, while the group of elderly accompanied by grandchildren showed improvement only on the mood scale. These results suggest several implications for the area of music therapy and music education, especially with regard to improving the healthy elderly's self-esteem, depression, and mood as components of successful aging.

With the increased elderly population in the United States (U.S.) there are growing concerns about their healthcare in the U.S. health industry. In the U.S., approximately 35 million people were at least 65 years old in the year 2000, and the numbers are expected to rise to an estimated 71 million by the year 2030 (Vincent et al., 2010). This growth, both in number and in percentage of the overall population, demands an active and knowledgeable approach by the healthcare community in order to maintain and enhance quality of life for the elderly. It also underlines the necessity of providing cost-effective alternatives to nursing home care, as the need for assisted living residences is expected to rise from 15 to 20% during this time period (Cummings, 2003).

Because of the rising elderly population, there has been increased interest in supporting the quality of life of elderly people through nursing care, social services, medicine, and other areas of health care. As early as 2002, Coffman pointed out that music educators and researchers should focus on effective ways to improve the quality of life for the elderly. Recently researchers have also indicated that as the life span in America has increased, there is a greater chance of facing healthcare problems such as the loss of physiological, psychological, and social functions (Chan et al., 2006; Chachamovich et al., 2008). Thus, it is important to establish programs that contribute to the quality of life of the elderly by involving them in daily living activities, physical exercise, and social interactions that provide healthy emotional outlets.

With regard to the music activities, some empirical research has emphasized the importance of quality of life for older people, investigating its aspects in several dimensions: psychological well-being, good physical functioning, relationships with others, health, and social activity (Bowling, 1995; Browne et al., 1994). Some music therapy studies have suggested that reminiscence therapy can help people adapt, and improve self-esteem, socialization, and life satisfaction and reduce sadness and depression by recalling and reviewing past experiences, feelings, and thoughts (Chao, 2006; Clair, 1996; Hays & Minichiello, 2005; Wang, 2004). In particular, such approaches have been examined for quality of life in the healthy elderly by having them participate in different music activity approaches (Bruhn, 2002; Coffman, 2002; Cohen et al., 2002).

Enhancing the quality of life for the elderly, however, is not built upon a single concept but rather a multidimensional concept that defines health as "a state of complete physical, mental, and social well-being, and not merely an absence of disease" (Constitution of World Health Organization, 2006, p. 2). In other words, the quality of life is defined in several ways. First, it has complex and multi-dimensional aspects comprising four domains: well-being, functional status, socioeconomic status, and self-esteem. Self-esteem refers to the degree to which people feel positive or negative about themselves and their self-acceptance, self-respect, and self-worth. In addition, self-esteem is the principal aspect of psychological well-being and is essential for successful and satisfying aging (Rosenberg, 1979). A positive experience of aging leads to higher subjective levels of well-being including self-esteem, and better physical functioning in later life (Wurm, Tesch-Römer, & Tomasik, 2007).

Depression is one of the most common mental illnesses in older adults. Twenty percent of older American adults experienced symptoms of depression in 2004 (American Psychological Association, 2005). Depression can result in the decline of physical function, severe emotional pain, and risk of death in the elderly population. In addition, depression affects the level of functional activity of daily living skills, cognitive functioning, life satisfaction, psychosocial well being and quality of life (Mosher-Ashley & Lemay, 2001; Ranzijn, Keeves, Luszcz, & Feather, 1998; Steffens, 2009; Rogers, 1999; Wong, Heiby, Kameoka, & Dubanoski, 1999).

Quality of life is closely related to "successful aging" and "active aging." The concept of active aging has factors in common with that of successful aging (Bowling, 1993, 2006, 2007). Rowe and Kahn (1997) identified three main components of successful aging: absence of and low risk of disease, a high level of psychological and physical function, and continued engagement in hobbies and other interests in order to maintain activities, and life satisfaction. Thus, successful aging is enhanced from the outset by using music activities in order to retain social, mental, and physical health.

Active aging refers to "the ability to be physically active," and to "participate in social, economic, cultural, spiritual, and civic affairs" (WHO, 2002, p. 12). Thus, successful aging and quality of life can be considered as "states of being." The concepts of active aging overlap with the most important components of successful aging and quality of life in order to enhance quality of life and mental and physical well-being (Bowling, 2008;Walker, 2002).

There is a considerable amount of literature concerning the benefits of music for the successful and active aging among elderly people. Listening to music is one of the most common leisure activities among the elderly and can be a frequent source of positive emotions. It affects the emotional states of older people and can be used as an aid to regulate mood, self-esteem, and group cohesion (Laukka, 2007). For example, several studies have used music therapy as a treatment intervention for elderly people with depression. (Ashida, 2000; Suzuki, 1998). Furthermore, choral music for healthy older adults can create an enjoyable environment through social interaction and cooperation in cross-generational choirs with college students (Darrow, Johnson, & Ollenberger,1994). Ernst and Emmons (1992) have suggested that participation in bands may also improve the social, mental, and physical functioning of seniors as well. These types of musical activities attribute positive social interaction as one of the most important elements of "productive and successful aging" for elderly people as they need and want to join in life enhancing activities and experience friendship and closeness to others (Bruhn, 2002).

Relatively few music educators and music therapists, however, have examined the effects of music on the healthy elderly, even though they can participate in various active and passive music activities such as choir, wind bands, piano clubs, and music therapy groups, which will contribute to their successful aging and psychological well-being. Clair (1996) found that music performance activities and general participation in music making with others may meet some basic human needs of the elderly and assist in maintaining and/or improving their psychological and physical well-being. Bruhn (2002) also argued that including choral singing in music therapy activities for elderly people improves their daily living skills and social interactions.

Taking the previous literature into account, the purpose of this research is: 1) To evaluate the effects of a combined music therapy and music education intervention on healthy elderly people, for the three dimensions of quality of life (overall quality of life, self-esteem, and depression) and mood, and 2) compare the effect of a combined music therapy and music education intervention among different elderly groups (the healthy elderly unaccompanied to treatment sessions, and healthy elderly accompanied by grandchildren (ages five to seven) during treatment sessions, and a control group receiving no treatment). In doing so, this study identifies and recommends the optimal music therapy and music education strategies for healthy elderly people on a long-term service basis.

To operationally investigate the purpose of this study, the following hypotheses were formulated:

Hypotheses

H1: There will be significant differences between three different experimental groups (music for the healthy elderly unaccompanied to treatment sessions, music for the healthy elderly accompanied by grandchildren (ages five to seven), and a no-treatment control group), based on the pretest to posttest gain scores on the overall quality of life, self-esteem, and levels of depression questionnaires.

H2: There will be a significant improvement for the first experimental group (music for unaccompanied healthy elderly) based on the pretest to posttest gain scores on overall quality of life, self-esteem, and depression questionnaires.

H3: There will be significant improvement for the second experimental group (music for the healthy elderly accompanied by grandchildren (ages five to seven) based on the pretest to posttest gain scores on overall quality of life, self-esteem, and depression questionnaires.

H4: There will be no significant improvement for the control group (no music intervention) based on the pretest to posttest gain scores on the overall quality of life, self-esteem, and depression questionnaires.

H5: There will be significant differences between the unaccompanied healthy elderly without grandchildren and the healthy elderly with grandchildren (ages five to seven) groups based on their pretest to posttest gain scores on the mood questionnaire.

H6: There will be a significant improvement for the first experimental condition (music group for elderly) based on the pretest to posttest gain scores on the mood questionnaire.

H7: There will be significant improvement for the second experimental group (the healthy elderly with grandchildren (ages five to seven) based on the pretest to posttest gain scores on the mood questionnaire.

METHOD

Subjects

The participants in this study were 41 volunteers, ranging in age from 55 to 70 years, from local churches in a small town in the southeastern United States. The participants were enrolled in one of three experimental groups. Eighteen individuals participated in the healthy elderly only experimental condition, six individuals participated in the healthy elderly accompanied by grandchildren (ages five to seven) experimental condition, and seventeen individuals participated in the no-treatment control condition. Each participant in the study met the following criteria: 1) the participant lived independently, as a single person or with a spouse in the community, rather than in a nursing home or assisted living center, and 2) none of the participants had cognitive impairment or dementia.

Instruments

The following evaluation instruments were used in this study: (1) a demographic questionnaire was used to gather information about the participants gender, age, and level of independent living; (2) the World Health Organization's Quality of Life questionnaire (WHOQOL-BREF) (World Health Organization, 1998), was used to evaluate the overall physical health, psychological well-being, social relationships, and environmental health of the participants; it includes 26 items (World Health Organization, 1998); (3) the Rosenberg Self-Esteem Survey, was used to evaluate the global self-esteem of the participants (Rosenberg, 1965); it has 10 items, which uses a Likert scale for simplicity of participants' responses. The Likert responses on the RSE ranged from 1 (completely agree) to 4 (completely disagree); (4) the Geriatric Depression Scale (Short Form; Sheikh & Yesavage,1986), was used to assess the participants' level of depression; it has 15 items, which is a short form of the GDS that originally included 30 items, and; (5) the Mood questionnaire, was used to assess the participants' state of mood. The research participants were asked to respond to the questions on the Mood questionnaire before and after each music therapy/music education intervention. The mood questionnaire, created by the researchers, was used to assess any changes in each participant's mood due to the music therapy/music education intervention (Eaglely & Chaiken 1993).

Procedure

For each experimental group, the protocol for music therapy and music education was as follows:

1. *Experimental condition unaccompanied healthy elderly*: This group met for 1 hour each week for six weeks to practice preferred songs that represented different styles of music such as religious music, American popular music, and/or patriotic music to engage the participants in active and passive music activities and choral singing.

2. *Experimental condition of healthy elderly accompanied by grandchildren* (ages five to seven)*:* The same protocol that was implemented for the unaccompanied healthy elderly experimental condition was implemented for the healthy elderly accompanied by grandchildren experimental condition.

The researcher, however, had to customize the music and songs used in the unaccompanied healthy elderly condition in order to meet the grandchildren's musical preferences.

3. *Control group*: Among the total of 41 participants in this research, 17 participants

were randomly assigned to the control group, which received no music therapy or music education intervention.

The research participants in all three experimental conditions were assessed before the first music therapy or music education session and again, after the last music therapy or music education session on the dependent variables of quality of life, self-esteem, and depression. The researcher administered the music therapy and music education experimental conditions for a period of six weeks. The participants' mood was assessed with the mood questionnaire, however, before and after each music therapy and education session. The mood assessment was conducted in order to assess to what extent each participant's mood may have changed from the beginning to the end of each music therapy/music education session.

RESULTS

Participant's Demographic Information

The majority of all participants (78%) were more than 60 years old (M = 66.02, SD = 6.27). More than 70% of all participants lived with their spouse and were financially self-sufficient, as well as highly educated (bachelor's or higher degrees). Their average frequency of doctor or hospital visits per month was 1.75 (*SD* = 1.27), but most of visits were for regularly scheduled medical checkups. Based on these facts, the subjects in this study sufficiently represented the characteristics of the "healthy elderly," who were self-sufficient, well-educated, and living with their spouse.

Hypotheses Testing

The hypotheses predicted that the combined music therapy/education intervention would yield significantly different effects between the three experimental conditions (the unaccompanied healthy elderly, healthy elderly with children, and no-treatment control groups) on the dependent variables quality of life, self-esteem, depression, and mood. In addition, the hypotheses also predicted that the music therapy/music education intervention would have positive effects on both the unaccompanied healthy elderly and healthy elderly with grandchildren groups from pretest to posttest. In other words, the pretest to posttest gain scores of individuals within each group were analyzed to determine if they had made significant progress from pretest to posttest on the dependent variables.

Table 1: Mean and Univariate F-value Differences on Dependent Variables between Experimental Groups

Experimental Conditions	Self-esteem	Depression	QOL-physical health	QOL-psychological health	QOL-social relationship	QOL-environment
Music group for unaccompanied elderly	3.22 (4.01)	1.22 (2.37)	.13 (.36)	.11 (.39)	.09 (.65)	.04 (.39)
Music group for elderly with grandchildren	.83 (2.99)	-.33 (.82)	.14 (.29)	.03 (.13)	.25 (.27)	.04 (.18)
Control group	.00 (1.70)	.18 (.88)	.05 (.37)	-.04 (.31)	.18 (.58)	.13 (.25)
Univariate F- Value	4.92*	2.60#	.26	.90	.20	.19

Note:
QOL = Quality of Life, Univariate d.f. = 2/38, #p<.10, *p<.05

An Analysis of Variance (ANOVA) statistical test was used to compare the mean gain scores between the three experimental groups on Hypothesis 1. Table 1 represents the participants' gain scores on the quality of life, self-esteem, and levels of depression. Regarding the mean differences among the experimental conditions, the unaccompanied healthy elderly group had higher gain scores than the healthy elderly with grandchildren group across the dependent variables. Multiple one-way ANOVA results, however, indicated a significant main effect for self-esteem only (F (2, 38) = 4.92, p < .05). Bonferroni post hoc comparisons for the dependent self-esteem variable was only significant for the no treatment control group (Δ = 3.22, p < .001), but not for the other comparisons

Table 2: Comparisons of Pretest and Posttest-Gain Scores on the Dependent Variables Within Experimental Group

Groups	Dependent Variable	Mean of Pre- and Post-test Gain Score	T-value
Healthy Elderly	Self-esteem	3.22	3.41**
	Depression	-1.22	-2.19*
	QOL - physical health	0.13	1.51
	QOL - psychological health	0.11	1.22
	QOL - social relationships	0.09	.61
	QOL - environment	0.04	.45
Healthy Elderly with Children	Self-esteem	0.83	0.68
	Depression	0.33	1.00
	QOL - physical health	0.14	1.22
	QOL - psychological health	0.11	0.54
	QOL - social relationship	0.25	2.24#
	QOL - environment	0.13	1.23
Control Group	Self-esteem	0.00	0.00
	Depression	-0.18	-0.82
	QOL - physical health	0.50	0.56
	QOL - psychological health	-0.39	-0.52
	QOL - social relationship	0.18	1.26
	QOL - environment	0.05	1.00

Note:
QOL = Quality of Life, #p<.10, *p<.05, **p<.01

($p > .05$). Thus, the results for H1 did not favor the healthy elderly groups in comparison to the control group on the self-esteem.

H2, H3, and H4 predicted that the music therapy/music education intervention would yield significant effects for both the unaccompanied healthy elderly and healthy elderly with children groups while no difference would be observed for the control group. Paired Sample T-tests were conducted to compare the participants' pretest and posttest gain scores on each dependent variable. As shown in Table 2, for the unaccompanied healthy elderly group, the difference between the pretest and posttest gain score means was significant for self-esteem (Δ = 3.22, t = 3.41, $p < .01$) and depression (Δ = -1.22, t = -2.19, $p < .05$). Therefore, the music therapy/music education intervention resulted in significant improvements on the participants' self-esteem and depression for this group. Thus, H2 was partially supported.

For both the healthy elderly group with children and control group, however, the difference from pretest to posttest gain score means was not statistically significant across any of the dependent variables in these two groups ($p > .05$). Thus, H3 and H4 were not supported.

Table 3: Comparisons between Pretest and Posttest Gain Scores on Mood

Group	Timing of Measurement	Mean of Gain Scores	Mean Difference	T-value
Healthy Elderly	Pretest	19.33 (3.69)	5.78	8.17***
	Posttest	25.11 (4.82)		
Healthy Elderly with Children	Pretest	14.00 (4.56)	1.17	2.91*
	Posttest	15.17 (4.36)		

Note:
*p<.05, ***p<.001

Meanwhile, H5 proposed that the gain score means of the dependent variables would be significantly different between the unaccompanied healthy elderly group and the healthy elderly with grandchildren group. An Independent Samples T-test was conducted to compare the gain score means on mood between the two experimental groups. As Table 3 indicates, the gain scores of the unaccompanied health elderly group were significantly greater than that of the health elderly group with grandchildren (Δ = 4.61, t = 5.67, $p < .001$). Thus, H5 was fully supported.

Furthermore, H6 and H7 predicted that the music therapy/music education intervention would yield significantly positive effects on mood scores for both the unaccompanied healthy elderly and the healthy elderly with grandchildren groups. As shown in Table 3, Paired Sample T-tests revealed that the difference between pretest to posttest mean scores was significant for both the unaccompanied healthy elderly and the healthy elderly with grandchildren groups (unaccompanied healthy elderly group: Δ = 5.78, t = 8.17, $p < .001$; healthy elderly group with grandchildren: Δ = 1.17, t = 2.91, $p < .05$). Therefore, implementation of the music therapy/music education intervention resulted in significant improvement of mood for both treatment groups. Thus, H6 and H7 were fully supported.

DISCUSSION

Based on the results of this study, the findings of this research provide several important implications for the participation of healthy elderly people in music therapy and music education activities. Perhaps the most important finding of this research was that the use of a combined music therapy and music education experimental condition was most beneficial for the unaccompanied healthy elderly group, especially with regard to mood, self-esteem, and depression. Although the music therapy and music education condition increased the levels of mood for both of the treatment groups, the improvement in levels of self-esteem and depression were significant only for the unaccompanied healthy elderly only group.

Considering that self-esteem and depression represent individuals' overall health and quality of life (see Eagly & Chaiken, 1993; Clements-Cortes, 2014), an interpretation of these results may be that the increased levels of mood in the healthy-elderly may be related to the levels of self-esteem and depression of participants. Thus it is interesting to note in this study that the unaccompanied healthy elderly group achieved improvement across the dependent variables of self-esteem, depression, and mood.

The results in this study indicated that the music therapy/music education sessions were not as effective for the healthy elderly accompanied by grandchildren (ages five to seven) group in comparison to the unaccompanied healthy elderly group. Except for improvement in the mood variable, there were no significant beneficial effects on self-esteem, depression, and quality of life for the healthy elderly with grandchildren group. Thus, the inclusion of grandchildren with the healthy elderly in a combined music therapy and music education experimental condition is not recommended as a result of this study. Indeed, previous studies have indicated that grandparents may experience physical and emotional hardships while taking care of grandchildren (Minkler et al., 1997; Szinovacz et al., 1999; Strawbridge et al., 1997). One plausible explanation may be that it is difficult for grandparents and grandchildren to maintain their concentration on music tasks during a combined music therapy and music education intervention.

This research study makes a unique contribution to the current music therapy research literature in two ways. First, this experiment recruited only healthy elderly individuals as research participants, while previous studies recruited research participants with a focus on elderly participants' specific diseases (Clements-Cortes, 2014; Choi et al., 2008). This research, however, suggests that the interventions of music therapy and music education may be useful as life management strategies, which may assist the healthy elderly in maintaining a successful aging process through improved levels of mood, self-esteem, and depression.

Moreover, the second unique contribution of this study is the combined use of a music therapy and music education intervention in efforts to maintain and/or improve the quality of life for healthy elderly participants. Music therapy and music education strategies were used both as a therapeutic and an educational intervention for the healthy elderly in this research. Although previous studies also employed different music therapy interventions (Choi et al., 2008; Clements-Cortes, 2014), few studies attempted to include a music education experimental condition in their music therapy intervention programs.

In this study, however, the music

therapy and music education experimental condition had a significantly positive effect on dependent variables that may be related to the attitudes of elderly participants with regard to successful aging. Indeed, during each session, all participants had opportunities to explore their inner states of mind and to review their own lives through creative music therapy and education approaches, such as lyric analysis, song discussion, instrument playing, and choral singing in a group setting.

In this study, the music therapy/music education experimental condition had a significantly positive effect on dependent variables that may be related to the attitudes of elderly participants with regard to successful aging. After each music session was finished, most of the participants commented on their love of playing instruments and choral singing. The participants indicated that playing instruments (music therapy) and choral singing activities (music education) were the activities that best affected their mood. In fact, the participants listed the following adjectives as descriptive of moods that were associated with playing instruments and choral singing activities: pleasure, delight, inspirational, fun, comfort, and meaningful. Many participants also mentioned that they loved the choral singing activities because they had opportunities to recall significant memories and moments with loved ones. Thus, the music intervention including music therapy and music education together was useful in improving mood, self-esteem, and depression of the healthy elderly as these interventions provided different psychological benefits such as happiness and life satisfaction (also see Menec, 2003).

In conclusion, this research suggests that music therapy and music education for the healthy elderly may improve their levels of mood, self-esteem, and depression. In addition, music therapy and music education interventions are especially suggested for unaccompanied healthy elderly participants. Regardless of the therapeutic or educational intent of such interventions, music can help the healthy elderly to express and share their emotions and feelings through listening, playing, and choral singing, which may in turn contribute to the enhancement of their self-esteem, depression, and mood, as components of successful aging. Thus, being engaged in music activities can help the healthy elderly by promoting their mental health and wellness through their music experiences with other people.

References

American Psychiatric Association. (2005). *Diagnostic and statistical manual of mental disorders* (5th ed.), Washington, DC.

Ashida, S. (2000). The effect of reminiscence music therapy sessions on changes in depressive symptoms in elderly persons with dementia. *Journal of Music Therapy,* 37,170–182.

Bowling, A. (1993). The concepts of successful and positive aging. *Family Practice, 10,* 449–453.

Bowling, A. (2006). Successful ageing from older people's perspectives. Results from a British survey of ageing. *European Journal of Ageing, 3,* 123–136.

Bowling, A. (2007). Aspirations for older age in the 21st century: What is successful ageing?. *International Journal of Aging and Human Development, 64,* 263–297.

Bowling, A. (2008). Enhancing later life: How older people perceive active ageing?. *Aging and Mental Health. 12*(3). 293-301.

Bowling, A. (2011). Psychological approach to successful ageing predicts future quality of life in older adults. *Health and Quality of life Outcomes, 9* (1), 13-22.

Browne, J. P., O'Boyle, C. A., McGee, H. M., Joyce, C. R. B., McDonald, N. J., O'Malley K., & Hiltbrunner, B. (1994). Individual quality of life in the healthy elderly. *Quality of Life Research, 3*(4), 235-244.

Bruhn, H. (2002). Musical development of elderly people. *Psychomusicology, 18,* 59-75.

Chachamovich, E., Fleck, M., Laidlaw, K., & Power, M. (2008). Impact of major depression and subsyndromal symptoms on Quality of Life and attitudes toward aging in an international sample of older adults. *The Gerontologist, 48*(5), 593-602.

Chao, S., Liu, H., Wu, C., Jin, S., Chu, T., Huang, T & Clark, M. J. (2006). The Effects of Group Reminiscence Therapy on Depression, Self Esteem, and Life Satisfaction of Elderly Nursing Home Residents. *Journal of Nursing Research 14(1), 36-45.*

Choi, A., Lee, M., & Lim, H. (2008). Effects of Group Music Intervention on Depression, Anxiety, and Relationships in Psychiatric Patients: A Pilot Study. *The Journal of Alternative and Complementary Medicine.14*(5), 567–570.

Chan, S, W., Chien, W., Thomson, D, R., Chiu, H, F, K., & Lam, L. (2006). Quality of life measures for depressed and non-depressed Chinese older people. *International Journal of Geriatric Psychiatry, 21,*1086-1092.

Clair, A. (1996). *Therapeutic Uses of Music with Older People*. Baltimore, MD: Health Professionals Press.

Clements-Cortes, A. (2004). Buddy's glee club two: Choral singing for older adults. *Canadian Journal of Music Therapy*. *20* (1), 85-109.

Coffman, D. D. (2002). Music and quality of life in older adults. *Psychomusicology 18*(1-2), 76-88.

Cohen, A., Bailey, B., & Nilsson, T. (2002). The importance of music to seniors. *Psychomusicology.* 18, 89-102.

Cummings, S. M. (2003). Impact of facility type on assisted living facilities' response to residents' mental health needs. *Journal of human behavior in the social environment, 8*(1), 25-41.

Darrow , A. A., Johnson, C. M., & Ollenberger, T. (1994). The effect of participation in an intergenerational choir on teens' and older persons' cross-age attitudes. *Journal of Music Therapy, 31,* 119-134.

Eagly, A. H., & Chaiken, S. (1993). *The psychology of attitudes*. Harcourt Brace Jovanovich College Publishers.

Ernst, R. E., & Emmons, S. (1992). New horizons for senior adults. *Music Educators Journal, 79*(4), 30-34.

Hays, T., & Minichiello, V. (2005). The meaning of music in the lives of older people: a qualitative study. *Psychology of Music*, 33(4), 437-45.

Laukka, P. (2007). Uses of music and psychological well-being among the elderly. *Journal of happiness studies, 8*, 215-24.

Menec, V. H. (2003). The relation between everyday activities and successful aging: A 6-year longitudinal study. *Journal of Gerontology. 58* (2). 74-82

Minkler, M., Fuller-Thomson, E. Miller, D., & Driver, D. (1997). Depression in Grandparents Raising Grandchildren: Results of a National Longitudinal Study. *Archives of Family Medicine 6*(5), 445-52

Mosher-Ashley, P. M., & Lemay, E. P. (2001). Improving residents' life satisfaction. *Nursing Home, 50*(5), 52-54.

Rogers, A. (1999). Factors associated with depression and low life satisfaction in the low-income, frail elderly. *Journal of Gerontological Social Work, 31*(1/2), 167-194.

Ranzijn, R., Keeves, J., Luszcz, M., & Feather, N. T. (1998). The role of self-perceived usefulness and competence in the self-esteem of elderly adults: Confirmatory factor analyses of the Bachman revision of Rosenberg's Self-Esteem Scale. *Journal of Gerontology Series B: Psychological Science & Social Science, 53B*(2), 96-104.

Rosenberg, M. (1965). Society and the adolescent self-image. Princeton, NJ: Princeton University Press.

Rosenberg, M. (1979). *Conceiving the self*. New York: Basic Books.

Rowe, J. W., Kahn, R. L. (1997). Successful aging. *The Gerontologist*, 37,433-440.

Stefens, D. (2009). A multiplicity of approaches to characterize geriatric depression and its outcomes. *Current Opinion in Psychiatry*, *22*(6), 522–526.

Strawbridge, W. J., Wallhagen, M. I., Shema, S. J., & Kaplan, G. A. (1997). New Burdens or more of the same? Comparing grandparent, spouse, and adult-child caregivers.

Gerontologist 37 (4), 505–10.

Suzuki, A. I. (1998). The effects of music therapy on mood and congruent memory of elderly adults with depressed symptoms. *Music Therapy Perspectives*, *16*, 75–80.

Szinovacz, M. E., DeViney, S & Atkinson, M. P. (1999) Effects of surrogate parenting on grandparents's well- being. *The Journal of Gerontology*. *54B*(6), 376-388

Vincent, G. K., & Velkoff, V. A. (2010). *The next four decades the older population in the United States: 2010 to 2050*. Retrieved from://www.census.gov

Walker, A. (2002). The evolving meaning of retirement. A strategy for active ageing. *International Social Security Review, 55*, 121–139.

Wang, J. (2004). The comparative effectiveness among institutionalized and non-institutionalized elderly people in Taiwan of Reminiscence Therapy as a Psychological, Measure. *Journal of Nursing Research, 12*(3), 237-244.

Wong, S. S., Heiby, E. M., Kameoka, V. A., & Dubanoski, J. P. (1999). Perceived control, self-reinforcement, and depression among Asian American and Caucasian American elders. *The Journal of Applied Gerontology, 18*(1), 46-62.

The WHOQOL group. (1998) Development of the World Health Organization WHOQOL-BREF Quality of Life Assessment. *Psychological Medicine*. *28*. 551–558.

World Health Organization. (2002). Active Ageing: A Policy Framework. World Health Organization, Geneva.

World Health Organization. (2006). Constitution of the World Health Organization. Basic Documents, Forty-fifth edition, Supplement. Retrieved August 11, 2012, from http://www.who.int/governance/eb/who_constitution_en.pdf.

Wurm, S., Tesch-Römer, C., & Tomasik, M. (2007). Longitudinal finding aging-related cognitions, control beliefs, and health in later life. *The Journals of Gerontology, Series B: Psychological Sciences and Social Sciences*, *62*, 156–164.

The Drum Set and Cerebral Palsy: Motor and Affective Psychological Development

Kaleb A. Santana Ramos & Emma Rodríguez Suárez

Conservatorio de Música de Puerto Rico

Abstract

Cerebral Palsy (CP) is a condition caused by an injury to the immature brain that can occur before childbirth until the age of three. This condition is non progressive and defined according to the muscular tone (high and/or low) and the area of the body affected. CP affects motor skills and creates limitations in movement. Moreover, CP could cause different brain and cognitive disorders such as learning disabilities, attention deficit, emotional impairments, behavioral problems, hydrocephaly, epilepsy, and visual impairments, among others. Due to the motor impairment and brain deficiencies that can limit the acquisition of proper skills it is difficult for a person with cerebral palsy to learn to play a musical instrument such as the drum set. The lack of data about the changes in motor skills and psychological affect impacted by learning such an instrument was the central purpose of this research. A qualitative, longitudinal case study approach was used to study the changes in fine and gross motor development as well as the changes in psychological affect in a student diagnosed with cerebral palsy who was musically educated on the drum set during a two and a half year period. Three methodological instruments were used to obtain data: 1) observations extracted from a pedagogical diary, 2) digital video recordings, and 3) interviews to the student and the student's mother. At the end of the analysis changes were observed in the following aspects studied: a significant increase in the speed on his general performance, from 118 to 162 (quarter note) and, an increase in endurance from four minutes and 13 seconds to 45 minutes of continued performance (motor), as well as significant changes in relation to the music, music education, and the student's future in the field of music (affective psychological).

Cerebral Palsy (CP) is caused by an injury to the immature brain, between fetus/child development stages and three years old. The manifestations are non-progressive and it is classified by the muscular tone (high and/or low), and the area of the body that is affected. Other related conditions with this diagnosis are learning disabilities, hydrocephalus, epilepsy, visual disabilities and behavioral diagnosis among others.

The main participant of this case study, Carlos Sebastián Egea, is a 21-year-old student with CP, with a specific diagnosis of spastic (high muscular tone), hemiplegic (right side of the body affected), epilepsy, strabismus, learning disabilities and oppositional defiant disorder (ODD). To aide him in his movement limitations he has undergone the following corrective surgeries: eye alignment (right eye), and tendon release (right arm and both legs). Furthermore, to minimize his epilepsy convulsions he has also undergone "vagus nerve stimulation therapy", a kind of pacemaker that sends mild pulses of electrical energy to the brain via the vagus nerve.

The need for this research study is based on the lack of data related to changes in motor skills (fine and gross) and affective psychological development on a student diagnosed with CP, specifically related to playing the drum set and its the impact when given individual lessons, in Puerto Rico (PR).

Research Questions

The following are the research questions that have led this study:

1) Which are the motor changes (fine and gross) that can be observed in a student with cerebral palsy after two and a half years of individual drum set lessons?
2) Which are the affective psychological developmental changes that can be observed in a student with cerebral palsy after two and a half years of individual drum set lessons?

Method

In this qualitative, longitudinal case study the teacher is also the researcher. There are two participants, the student and the student's mother. The instruments used to collect the data were three: observations (written in a pedagogical journal), digital videos recordings, and interviews (post) to both participants. Documented observations were made throughout the term of the two and a half year period.

Results and Conclusions

The data collected during the two and a half years of this research suggests changes in both areas studied, motor (fine and gross) and affective psychological. These changes (motor and affective psychological) suggest a direct link between them and the individual drum set lessons.

Motor. As an example of the data collected, Carlos was able to master 12 out of the 19 (63%) technical exercises that he was presented with, and showed increase in his performance tempo in the drum set (in general) from 118 to 162 quarter note (metronome).

Affective psychological. The student's body expression suggested comfort and relaxation while playing the drum set, which lacked at the beginning of the research. His verbal expressions demonstrated interest in continuing to improve his performance and to expand his repertoire. The student also showed motivation and pleasure in participating in live performances and externalized his desire to continue musical studies at the university level.

Implications. The three implications that emerged from this research are: time spent in music class, teaching strategies, and student's motivation. These came about as part of the learning process and the effort to make Carlos Sebastian progress.

Time spent in music class. Usually, students with special needs require more time and attention to complete the tasks require in class. The teacher has to invest additional time to ensure that the student shows understanding of the class material and makes progress. In a classroom situation where there are 15 to 20 additional students, the teacher has to make room to attend all types of students, with special needs and without, in order to fulfill teaching goals; even if it means teaching a student with special needs after school hours.

Teaching strategies. Teachers have to be effective on delivering information to students. They need to be creative on the use of teaching strategies in order to succeed and ensure the student's progress. For example, for Carlos Sebastian to understand the technical exercises, a new form of musical notation was created; a simplified version of performing the drum set while reading and counting rhythms at the same time. This new notation had two purposes: first, to help him learn and play new exercises and rhythms; and second, to prepare him to learn traditional music notation. Once these technical exercises were mastered, the next step was to apply them to the repertoire. This teaching strategy among others was used to help Carlos Sebastian learn new technical exercises and rhythms to help him progress on the drum set.

Student's Motivation. Motivation is the key to maintain students on the path to learning. Teachers need to listen to their students and be aware of the student's taste in music, so they can use it as a tool to keep them motivated. Selecting a piece of music or a song based on the student's taste would represent an advantage for the teacher. The music teacher can select a song that students would love to play and at the same time represents a technical challenge for them. This action would help the teacher fulfill five objectives. One: to keep students motivated by playing their favorite music; two: to make them progress through practice; three: to play their favorite music; four: to make students practice; and five: to make the teacher maintain control of the repertoire. Using student's favorite music, songs are selected under the teacher's guidance based on the student's preference to facilitate the learning process by motivating students to participate on the selection of the repertoire.

Student's motivation in itself is a loop. When students see themselves progressing, this keeps them motivated. In this study, helping Carlos Sebastian do things that he didn't know he could do gave him the opportunity to develop in something that he can learn and be good at. This sense of accomplishment kept him motivated and moving forward in learning and wanting to progress.

The implications described here are not written in a priority list; they interact with each other. They create a symbiotic cycle that generates a win-win situation for both, teachers and students.

Recommendations

Based on the implications emerged from this research, five recommendations are made. One: expose students to listen to different musical styles (teaching strategy); two: take students to attend live concerts to observe the performance of other musicians (investing additional time, teaching strategy, motivation tool); three: take students to visit educational musical institutions to interact with other teachers and students (investing additional time, teaching strategy, motivation tool); four: take students to participate in musical and instrumental workshops (investing additional time, teaching strategy, motivation tool); and five: use videos onYouTube.com to illustrate as a model and teaching tool (teaching strategy, motivation tool).

These implications and recommendations emerge from the results and conclusions, and the experience of the teacher/researcher from the process of teaching individual drum set lessons to a student diagnosed with CP through a period of two and a half years.

References

Arnau, J., & Bono, R. (2008). Estudios longitudinales. Modelos de diseño y análisis de medidas repetidas. *Escritos de Psicología, 2*(1) 32-41.

Dorneles, F., Freitas, J. C., Pavan, F., Kayenne, C., & Formiga, R. (2012). Correlation between neurofunctional profile and sensory-motor skills of children with cerebral palsy. *Journal of Human Growth and Development, 22*(2), 226-232.

Kara, O. K., Mutlu, A., Günel, M. K., & Livanelioğlu, A. (2012). Effect of acquisition ages of gross motor functions on functional motor impairment in children with cerebral palsy. *Turkish Archives of Pediatrics, 47*, 191-196. doi:10.4274/tpa.896

Kim, S. J., Kwak, E. E., Park, E. S., & Cho, S. R. (2012). Differential effects of rhythmic auditory stimulation and eurodevelopmental treatment/ Bobath on gait patterns in adults with cerebral palsy: A randomized controlled trial. *Clinical Rehabilitation, 26*(10), 904–914. doi: 10.1177/0269215511434648

Krigger, K. W. (2006). Cerebral Palsy: An overview. *American Family Physician, 73*, (1) 91-100.

Lacársel, J. (2003). Psicología de la música y emoción musical, *Educatio,* 20-21, 213-226. Extraído de: http://revistas.um.es/index.php/educatio/article/viewFile/138/122

López Estrada, R. E., & Deslauriers, J.-P. (2011, Junio). La entrevista cualitativa como técnica para la investigación en Trabajo Social. *Margen, 61*, 1-19.

Lucca Irizarry, N., & Berrios Rivera, R. (2003). *Investigación cualitativa fundamentos, diseños y estrategias*. San Juan, PR: Publicaciones Puertorriqueñas.

Mcferran, K. S., & Shoemark, H. (2013). How musical engagement promotes well-being in education contexts: The case of a young man with profound and multiple disabilities. *International Journal of Qualitative Studies on Health and Well-being, 8*, 1-14.

Morais, F. D., Freitas, J. C., Viana, F. P., & Formiga, C. K. M. R. (2012). Correlation between neurofunctional profile and sensory-motor skills of children with cerebral palsy. *Journal of Human Growth and Development, 22*(2), 226-232.

Panksepp, J., & Bernatzky, G., (2002). Emotional sounds and the brain: the neuro-affective foundations of musical appreciation, *Behavioral Processes, 60*, 133-155.

Reynoso, K. M. (2010). La educación musical y su impacto en el desarrollo. *Revista de Educación y Desarrollo, 1,* 53-60.

Rosenbaum, P., Paneth, N., Leviton, A., Goldstein, M., & Bax, M. (2006). A report: the definition and classification of cerebral palsy April 2006. *The Definition and Classification of Cerebral Palsy,* 8-14.

Soria, G., Duque, P., & García, J. M., (2011). Música y cerebro (II): Evidencias cerebrales del entrenamiento musical. *Revista Neurológica, 53*(12), 739-746.

St. George, J., Holbrook, A., & Cantwell, R. (2014). Affinity for music: A study of the role of emotion in musical instrument learning. *International journal of music education, 32*(3), 264-277.

Stoshljevikj, M. L., Fadilj N., Eminovikj, F. N., Nikikj, R. M., Gordana I., Achikj, G. I., &P acikj, S. R. (2008). Cerebral palsy and music achievement. *Journal of special education and 24 rehabilitation, 9*(3-4), 23-37.

Uttech, M. (2006). ¿Qué es la nvestigación-acción y qué es un maestro investigador? *Revista de Educacion, 8*.

Voorman, J. M., Dallmeijer, A. J., Knol, D. L., Lankhorst, G. J., & Becher, J. G. (2007).Prospective longitudinal study of gross motor function in children with cerebral palsy. *Archives of Physical Medicine and Rehabilitation, 88*, 871-876.

MUSIC IMPROVISATION SKILLS: THERAPEUTIC TOOLS FOR RESOURCE DEVELOPMENT IN CLINICAL PRACTICE

Lynn J. Saltiel, MSSW, LCSW, Certified EMDR Therapist
Hudson Psychotherapy and Consultation

Abstract

Mental health and allied professionals are often conduits for fostering hope and change in clients that present multiple complex clinical concerns within multiple practice settings. In this experiential workshop, participants will be introduced to universal music and movement improvisational structures and techniques for building individual resiliency, interpersonal communication skills, as well as creative self-empowerment. Relevant theoretical literature will be reviewed. Several case examples will also be included highlighting clinical applications from a women's therapeutic support group within the context of an urban outpatient community mental health center. This workshop will offer a humanistic philosophical approach to music making which holds the belief that innate music making is an available tool for everyone to enhance health and wellbeing. Improvising is creating spontaneously an experience in the moment. To improvise takes courage and authenticity. When we improvise, we face the unknown. This process can help one face their fears, developing courage and trust. Self-expression can lead to self-actualization and a sense of building mastery over one's abilities that can generalize to other life experiences leading to self-growth and empowerment. The skills that encompass sound therapeutic practice are mirrored in similar skills that make up shared musical practice such as deep listening to self and others, supporting, trusting, mirroring, reflecting, releasing judgment, being in the present moment, authenticity, attunement and connection. Additional psychological benefits include a greater sense of community and belonging through this shared experience.

Music melts all the separate parts of our bodies together- Anais Nin

Improvisation is a key musical skill. It is the creative foundation in composition development; an arena for interpersonal interaction and playfulness; an outlet for personal expression; and provides a useful mechanism for developing coping skills needed to navigate life's challenges.
Mental health and allied professionals are often conduits for fostering hope and change in clients that present multiple complex clinical concerns within multiple practice settings. While clinicians can draw upon any number of talk therapy techniques to help their clients, there are times when verbal interventions aren't helpful or words cannot be summoned. Music has been described as the "universal language." Improvisation as an exposure to non-verbal techniques can be broadening. Music can be used to address a myriad of clinical themes in a non-confrontational way. It can offer our clients a non-verbal way of expressing themselves and communicating their needs.
Learning some basic universal music and movement improvisational structures and techniques can enhance building individual resiliency, interpersonal communication skills, as well as creative self-empowerment. These tools can be viewed as a valuable avenue for helping unite a group, fostering camaraderie and improving social skills.
On an individual level, it can be a dynamic channel for developing personal insight into feelings and promoting critical examination of life issues, struggles and experiences. Knowledge, self-efficacy and emotional-regulation, for both individual and interpersonal skills, are needed in order to productively navigate daily life. Many client populations that mental health professionals encounter lack one or more of these key developmental facets.
Facilitation of music improvisational skill building and the metaphors for the ways individuals co-create interactions that are "more than the sum of the parts" can be employed within many practice settings.
From a humanistic philosophical approach to music making, this holds the belief that innate music making is an available tool for everyone to enhance health and wellbeing.
If we consider health and wellbeing to be viewed under the construct of "wellness", how do we then define this construct?
What is wellness? According to the *World Health Organization*, wellness consists of "...A state of complete physical, mental and social well-being and not merely the absence of disease or infirmity." An alternate definition comes from *The National Wellness Organization* which states it is "A conscious, self-directed and evolving process of achieving full potential." Finally, wellness can be viewed as "an active process of becoming aware of and making choices toward a healthy and fulfilling life" (University of California, Davis). (Note: Wellness definitions retrieved from internet on 4/5/15).
Improvising is creating spontaneously an experience in the moment. To improvise takes courage and authenticity. When we improvise, we face the unknown. This process can help one face their fears, developing courage and trust. Self-expression can lead to self-actualization and a sense of building mastery over one's abilities that can generalize to other life experiences leading to growth and self-empowerment.

The skills that encompass sound therapeutic practice are mirrored in similar skills that make up shared musical practice such as deep listening to self and others, supporting, trusting, mirroring, reflecting, releasing judgement, being in the present moment, authenticity, attunement, and connection. **Human connections** shape the **neural connections** from which the mind emerges. From a developmental and interpersonal biology perspective concur that at least some part of our early attachment and interpersonal experiences effect and help shape whom we later become as adults. By altering the activity and the **structure of the connections** between neurons, **experience directly shapes the circuits** responsible for memory, emotion and self-awareness (Siegel, 1999, 2007) Healthy attunement involves the parent's sensitivity to the child's signals and the collaborative communication between the two people's states of minds: the mutual influence of each person's state on each other (Siegel, 1999, 2007).

Attachment issues are about attunement. The ability for an individual to be self-reflective, compassionate and empathic enough to consider the question what is it like for you in this moment? Secure attachment requires an array of emotional skills: affective competence, feeling and dealing while still relating, ability to feel and process emotions in a dyad, flexible strategies in relational experiences of affect, ability to maintain integrity of self and other, able to tolerate separation and reunion without being overwhelmed, and the capacity of self and interpersonal regulation (Lalilotis, 2014).

Assessment Considerations-Some Typical Needs for Each Developmental Stage

Age	Some Typical Needs
0-1 years	Food, safety, nurturing, love, closeness, attunement, repair of relationship breaks, caretakers who can manage their emotions and meet their own needs
1-3 years	...*plus* Reassurance, unconditional acceptance, praise, respect, loving correction, play, freedom to develop self-control, stimulating exploration, consistency, structure
3-6 years	...*plus* Encouragement, approval for accomplishments, validation, opportunities to make things, opportunities to master motor skills, freedom to express emotions, help processing through strong emotions (i.e. grief, anger)
6-12 years	...*plus* Learning opportunities and intellectual challenges, physical challenges, freedom to develop and express opinions, success at problem-solving, support with social problems/challenges
12-18 years	...*plus* Freedom to begin separating from parents, encouragement and freedom to find a sense of identity, encouragement and support for making educational and occupational decisions

Part of the assessment process in considering utilizing music improvisation as a clinical resource tool is to identify the key developmental tasks that may have gotten arrested or may never have been fully developed over the course of an individual's life. Human development is seen as a series of phase specific tasks, the successful mastery of which leads to the emergence of more and more complex psychic structures. Psychopathology is often viewed as a failure to master expected developmental tasks or predicted structural achievements. Since each developmental stage builds on the next, this is an important assessment consideration.

For instance, the clinician can be alerted to what attachment and developmental needs may have been thwarted or not met and how the therapeutic process can best proceed. It is important to pay close attention to the early needs such as safety, trust, acceptance, validation, support, control/choice, connection, and ability to play. The chart provides a summary of some age typical needs for each developmental stage.

Music Improvisation is a way to facilitate play, creativity and self- discovery. It can create a space, or in the words of Donald Winnicott "a holding environment." Through the process, it can help the person begin to become aware and notice their experience in a more mindful way. Free music improvisation does not involve the thinking mind. It is actually the state of no-mind that provides a wonderful container into which obscure thoughts and feelings can be externalized and transformed through the symbolic language of music (Montello, 2002). The individual's unique authenticity may never have been seen or heard by the people closest to them. In a preverbal or very young child, there is a limited vocabulary and ability to articulate feelings and experiences. Memories are trapped. Musical improvisation is a way to develop or reclaim the voice of the self. Feelings can be released, explored and transformed. It is actually the state of no-mind that provides a wonderful container into which obscure thoughts and feelings can be externalized and transformed through the symbolic language of music (Montello, 2002). Music bypasses cortical functioning and goes directly to the emotional centers where the unexpressed or underdeveloped feelings are at the root of most relationship issues. By being completely in the moment, an individual is creating a musical snapshot, a musical self-statement.

Music and sound is subjective-there is no formula. Being together in music improvisation allows for just being, instead of having to do something "right." This allows for a whole new level of expression and sense of self. Sensitivity to cultural competency and diversity issues and how this influences an individual's relationship to music and sound is both astute and ethical practice. This includes an awareness of not only the dominant cultural heritage or the clinician's own heritage but to becoming culturally aware of the heritage of others, including "…. *artistic expressions"* of major client groups served (NASW, 2007). Exposure and a working knowledge and understanding to a wide range of nonwestern based musical traditions can be applied to clients with many diverse backgrounds and traditions. Resource development can be viewed in terms of the individual's coping abilities but it can also be viewed more broadly to include the concept of collaborating with others. Collaboration involves interpersonal communication, group process skills, and empathy. These are also key features in improvising musically with others. Groups can include families and other groups, such as therapeutic groups, teambuilding and community organizing. Psychological benefits include a greater sense of community and belonging through this shared experience, development of social skills, and interpersonal interaction. Of course, it is necessary to create a safe space for people within the group to feel comfortable enough to risk expressing themselves or parts of self to others. This can be quite a vulnerable place to be and the clinician/facilitator has a duty to be aware of this and guide the process accordingly. The goal of the process is not about the individual helping professional showing off their musical skills; it is about empowering people to discover their own inner music in an open, fun, and supportive environment. Observations in group dynamics include: characteristic roles of the individual members (re: leader, follower, scapegoat), on-going power struggles, flexibility of members to listen and respond to the others, communication difficulties, creative blocks and non-verbal communication style (i.e. eye contact, proximity of seating positions).

There are some basic yet powerful improvisation structures that enhance and can develop authentic individual music making or used to create music in ensembles or group with others. These techniques were developed, learned and inspired through the *Music for People* organization which promotes a humanistic approach to the way music is taught, performed, and appreciated.

The core philosophy asserts respect and the belief that the creative impulse is within everyone and it is vital to the process that individuals free themselves and let go of thoughts, judgement, get in the body and honor the child within (Oshinsky, 2008). Since there are many activities, structures and techniques to access from for music improvisation, a select sample in which this clinician has greater experience using in therapeutic practice. The case examples discussed highlighting these techniques are from a women's open-ended therapeutic support group within the context of an urban outpatient community mental health center. The description and information contained here are qualitative. The goals for this group are to provide mutual aid and support within the community, increasing self-esteem, and developing individual empowerment strategies. The group has been exposed to multiple forms of creative arts interventions including art, journal writing, yoga, meditation, movement and music improvisation. Additionally, the group has also participated in more traditional talk therapy, experiential and didactic approaches. Some of the participants are also attending individual psychotherapy, as well as some are also under the care of a treating psychiatrist. Some members are from the community and are not engaged in other clinical services. The demographics of the participants are multi-cultural and multiple ages, though the median age range is late 50's to mid-60's.

Most of the participants receive some form of public assistance currently, even if they had been previously employed. All the core group participants have a minimum of a high school education with some having college experience or a college degree. The participants' level of family and other social support is varied, however, the group members verbalized that the group serves as an important source of support and decreasing a sense of isolation. As far as

musical experience is concerned, all of the participants considered themselves to be novices, with little to no previous musical experience.
The following descriptors are examples of several improvisational techniques that can be used in clinical practice:

Babbling
Babbling is what babies do setting up their mouth and muscles for speech. Babies make the same sounds no matter what culture or language they originate from. Babbling is a universal language.
Most of us stop using babbling once we develop speech. As a music/vocal technique, babbling helps us to warm up all the mechanisms in our mouth. It can also be transferred to playing an instrument. It can be used as an icebreaker to help put people at ease and as a way to loosen up and let go of any critical thoughts or some inhibitions someone may have. In group, we have used this in the form of musical conversations and it is always interesting to see how people relate to one another using babbling and other non-verbal aspects of communicating.

One Quality Sound
When you listen to one note at a time, you are really listening to every sound you play. Simply, to create an improvised melody, begin with one note or one quality sound. Listen. This will lead to the next quality sound, creating a line or phrase. In a group, individuals can try this all together with first taking a breath in together and on the exhale all the members come up with their own individual note at the same time. The idea is NOT to try to match pitch but stay true to your individual sound. It is amazing to hear the beautiful harmonies that are created by this approach. "One quality sound" teaches to trust the unknown and to surrender in the moment, becoming one with the sound. As a group exercise, it connects the group. The collective one quality sound also provides for less individual exposure for individuals who may be more tentative.
This can be a way to start an improvisation with an ensemble (Miller, 2013, p.74).

Offer Your Silence
Silence is your friend, as per *Music for People*. It is fine to offer your silence if you are not sure what to play. Individually or playing with others, silence creates interest. Additionally, it is helpful to be aware that everyone does not have to be making a sound or playing all the time. In the women's group, this was a valuable lesson learned and during feedback it was noted that it is similar to having a conversation. If everyone talks at the same time, all the time, it is probably not a very fulfilling conversation! This relates to building social skills and communication skills with others.

Call and Response
When playing in a group or ensemble, if an individual hears something they like, it is good to imitate it, double it, or make a variation on this. This validates the other, shows that you are listening to others, and provides continuity. Call and response begins with one person creates a call and the other person or the rest of the group, echoes the call back. This technique works well in all kinds of group situations, this exercise gives everyone the opportunity to be a leader, and it is supportive and empowering to hear your phrase repeated back (Miller, 2013, p. 76)

Shadowing

This technique gives validation and connection with another. With a partner, one person is the leader, the other is the follower. The follower mirrors exactly what is seen and heard. Trade roles in following and leading and then find the wondrous place where no one is leading or following but both have come together. Shadowing can be an effective exercise in deep listening and intuition. It can also be a great diagnostic tool to process for those who may experience difficulty with those individuals in this activity.

Solo Over Drone

A drone is one long tone held. The drone is heard in music from many cultures around the world. This powerful sustained note is home. It is safe; it's consistent, as an accompaniment for a soloist it is very supporting. It is great to try this with a dyad where each person has a chance to switch both roles, a few times to experience the difference. A group can also hold a drone for a soloist who needs the support.

Ostinato

A repeated pattern is called an ostinato. There is power in repetition. It is grounding. To repeat a pattern either with the voice or an instrument, it requires focus to learn how to hold a pattern. For many individuals, they may be distracted or experience racing thoughts, thinking that there is a need to constantly do more. However, doing less often leaves more room for expansion (Miller, 2013, p.79). This can be an example of mindfulness skills training.

Conclusion

Most of the participants remarked in numerous sessions that they were distracted, agitated or anxious at the onset of the group and by the end; they felt more focused, grounded and less anxious or upset. A number of participants also remarked that the quality of the relationships, the connections experienced amongst the members, felt mores solid once music improvisation activities were introduced.

In conclusion, some of the participants in the women's group remarked that the music improvisation exercises also benefitted them to help remove other creative blocks they had been experiencing in another creative processes, such as writing or creating a piece of artwork. Were the participants in this group more open than other individuals through self-selection in a group process that had already been utilizing creative arts approaches? Would this account for the reported benefits? This clinician has participated in various kinds of clinical situations using the same techniques and structures; for instance, organizational teambuilding, and community improvised music sessions, couples therapy sessions, and individual therapy sessions. In all of these various modalities, consistent feedback was given on the development of greater self- awareness, observations on how communication styles and interpersonal relationships that were out of consciousness were brought to light through the use of music improvisation. The positive intention on the part of the clinician for the possibility of client change and the active participation of the client(s) has been measured to be key components in obtaining greater results in a number of wellness studies (Stevens, 2012).

All of these observations lead to further questions rather than answers. It is the hope of this clinician that further studies will be explored to greater understand the social-emotional health benefits of using music improvisation skills in clinical practice.Improvisation is fun, it is accessible, it is interactive, and the skills developed assist in broadening personal expression and freeing inner inhibitions for personal growth, health and wellbeing.

REFERENCES

Boyd Webb, N. (2004). Music Therapy to Help Traumatized Children and Caregivers, (chapter 9) in *Mass Trauma and Violence: Helping Families and Children Cope.* New York: The Guilford Press

Campbell, D. (2006). *Music-Physician for Times to Come.* Wheaton Il. Quest Books

Hart, Michael Anthony (February 2010). *Indigenous Worldviews, Knowledge, and Research: The Development of an Indigenous Research Paradigm.* Journal of Indigenous Voices Social Work, Volume 1, Issue 1.

Hull, A. (1998). *Drum, Circle, Spirit. Facilitating Human Potential Through Rhythm.* Gilsum, NH: White Cliffs Media.

Laliotis, D. (July 2014). "Going Deeper Into Personality and Character Structure Using EMDR Therapy. An EMDR Master Class. Co-sponsored by The EMDR Institute and Bender/Britt Seminars. Woodbridge, NJ.

Mannes, E. (2011). *The Power of Music: Pioneering Discoveries: the New Science of Song.* NY: Walker and Company.

Montello, L. (2002). *Essential Musical Intelligence. Using Musicas your path to healing, creativity, and radiant wholeness.* Wheaton Illinois: Quest Books.

Miller, L. (2013). *SpiritArts. Transformation through Creating Art, Music and Dance.* Phoenixville, Pa: Expressive Therapy Press.

National Association of Social Workers. (2007). *Indicators for the Achievement of the NASW Standards for Cultural Competence in Social Work Practice.* Retrieved from http://www.socialworkers.org/practice/standards/NASWCulturalStandardsIndicators2006.pdf Washington, DC: NASW Press.

Oshinsky, J. (2008 Edition). *Return to Child: Music for People's Guide to Improvising Music and Authentic Group Leadership. Philosophy, Games, and Techniques.* Goshen, Ct: Music for People.

Shapiro, F. (2012). *Getting Past Your Past-Take Control of Your Life with Self-Help Techniques from EMDR Therapy.* NY: Rodale.

Siegel, D. (1999, 2007). *The Developing Mind. How Relationships And The Brain Interact to Shape Who We Are.* New York: The Guilford Press.

Stevens, C. (2012). *Music Medicine: The Science and Spirit of Healing Yourself With Sound.* Boulder Co., Sounds True, Inc.

Study on Music and Sickle Cell Disease: Concept Report

Kristie Lipford, David Akombo, Wendy Thompson

Jackson State University

Abstract

There is evidence that shows that music listening has a positive effect on pain (Ghaffaripour et al., 2007; Linnemann et al., 2015; Mitchell & MacDonald, 2006; Pothoulaki et al., 2008). However, there is little research that focuses on the effects of music listening and pain perceived by those suffering from sickle cell disease (SCD). To date, there is one study that examines this connection (University Hospital Case Medical Center, 2014). Researchers at University Hospital Case Medical Center are conducting a clinical trial entitled "The Effects of a Single Music Therapy Session on the Pain of Adult Patients with Sickle Cell Disease: A Mixed Methods Study" which investigates how music listening influences patients mood, pain intensity, and pain relief (University Hospital Case Medical Center, 2014). Our proposed study is different from the research being conducted at Case Medical Center in that we plan to run an intervention for several months whereas the clinical trial at Case consists of one single session of music listening. Extended music listening will allow us to see medium term effects of music listening, something that is not feasible with just one session of music. To our knowledge, no study has examined this relationship medium-term, thus, this kind of investigation is greatly needed.

Hypotheses guiding the study

Based on previous studies that evidence the effectiveness of music on pain perception, we hypothesize that the music listening intervention will be negatively associated with chronic pain perception. We also hypothesize that the music listening intervention will be negatively associated with hospital visits, ER visits, and conventional medication use.

H_{1a} – Socioeconomic factors will be associated with the effectiveness of music listening on the pain perception.

H_2 – Study participants in the experimental group will have lower levels of chronic pain than study participants in the control group.

H_3 – Study participants in the experimental group will have less hospital visits than study participants in the control group.

H_4 – Study participants in the experimental group will have less emergency room visits than study participants in the control group.

H_5 – Study participants in the experimental group will have a lower frequency of conventional medication use than study participants in the control group.

Purpose of the Study:

There are very few publicly available datasets that sample a population of people affected by SCD. Moreover, contemporary research is just now starting to recognize the efficacy of music and other creative arts therapy in medicine. Past studies have shown that music is an effective medium to ease pain perception and decrease the threshold of pain (Ghaffaripour et al., 2007; Linnemann et al., 2015; Mitchell & MacDonald, 2006; Pothoulaki et al., 2008). Although music therapy as an established allied health profession has produced numerous empirical and qualitative data on increased frequency in medical treatments, there is a dearth of empirical research that supports the use of music on patients with sickle cell disease. In addition, many of these studies are few and findings are often contradictory. This research will be devoted to examining the effects of music on the pain perception of patients living with sickle cell disease.

REVIEW OF THE RELATED LITERATURE

Sickle cell disease (SCD) is categorized by episodes of severe pain that mainly affects the chest, back, and legs of diagnosed individuals (Thompson, 2014). Experiences of pain is the primary cause of hospitalization among SCD patients

(Marlowe & Chicella, 2002; Thompson, 2014). These hospitalizations are very costly and it is estimated that $475 million dollars are spent annually on treating SCD pain (Marlowe & Chicella, 2002). SCD pain treatment varies from patient to patient and the type of treatment received depends on pain severity (Okpala and Tawill, 2002; Thompson, 2014). Patients who experience acute pain typically are prescribed high dosage opioids which are injected several times a day. Those with less severe pain are often prescribed anti-inflammatory medications, such as ibuprofen, while others self-medicate with over the counter pain relievers (Marlowe & Chicella, 2002).

The one disadvantage with prescribing opioids is that many patients often become tolerant to the medication forcing rheumatologist to increase the prescribed dosage (Marlowe & Chicella, 2002). Other studies have commented on the reluctance of attending physicians due to the risk of substance abuse and addiction (Marlowe & Chicella, 2002). Other doctors are often reluctant, as well as SCD patients, due to the undesirable side effects of opioids (Marlowe & Chicella, 2002). The treatment of SCD pain is quite concerning and more research needs to be conducted to assess the effectiveness of non-opioids pain management.

The effects of music on pain have recently been examined, but very few studies have examined the effects of music on pain perception in patients living with sickle cell disease. This kind of investigation is greatly needed. The application of music on pain management and the control of pain has increased in the last two decades. Past studies have shown that music is an effective medium to ease pain perception and decrease the threshold of pain (Ghaffaripour et al., 2007; Linnemann et al., 2015; Mitchell & MacDonald, 2006; Pothoulaki et al., 2008). Although music therapy as an established allied health profession has produced numerous empirical and qualitative data on increased frequency in medical treatments, there is a dearth of empirical research that supports the use of music on patients with sickle cell disease. In addition, many of these studies are few and findings are often contradictory. This research will be devoted to examining the effects of music on pain perception on patients with sickle cell disease.

RESEARCH DESIGN & METHODS

This is a pilot study that will make use of qualitative and quantitative data to assess the effect of music listening on pain perception of sickle cell patients aged eighteen and older suffering from chronic pain. The proposed study period will run from January 2016 to April 2016. Inclusion criteria for this study will include sickle cell patients in the outpatient clinic over the age of 18 who experience chronic pain. Exclusion criteria for this study will exclude patients who are using any other alternative medicine aside from conventional medication to treat sickle cell pain. The predictor variable is music listening. This study has four outcome measurements:

1. chronic pain level (interval level measurement)
2. hospital visits (interval level measurements)
3. ER visits (interval level measurements)
4. medication usage (interval level measurements)

Also of interest are social patterning factors that may be associated with the intervention. Factors such as SES, age, race, gender, weight, tobacco smoking status, drug therapy, alternative medicine use, and co-morbidity diagnoses will be used to determine whether these factors have a relationship with the outcome variable.

Setting and Sample Participants:

The sample will consist of a single site recruitment of 50 patients who are scheduled to receive SCD treatment at an adult sickle cell clinic in Jackson, MS. The nursing staff and other relevant medical professionals at the clinic will assist in recruiting participants. Twenty five of the participants will serve as the experimental group, while the other 25 will serve as a clinical control group via chart evaluation. Recruitment fliers will also be posted in the waiting room of the clinic. Participants will be included in the study if they are over 18 years of age, have a diagnosis of SCD, and have a history of SCD derived pain. The principal investigator will seek full review and approval by the Institutional Review Board (IRB) at Jackson State University and University of Mississippi Medical Center (UMMC) to conduct this study.

The participants in the proposed study will represent the demographics of the UMMC Jackson Medical Mall Sickle Cell Clinic population. Population demographics of the clinic include majority female, low to middle income SES, predominantly African Americans. A large proportion of patients also are Medicaid and/or Medicare beneficiaries.

Instrumentation:

In addition to collecting, demographic data one instrument, Numeric Pain Rating Scale (NPRS), will be used to capture the pain perception of the participants (McCaffery et al , 1989). The Numeric Pain Rating Scale is an 11 point scale that ranges from 0-10, with 10 being the highest level of perceived pain (severe pain).

Procedures:

Once recruited, participants will be informed of the clinical trial and asked to sign Consent Form if they agree. The study will last 16 weeks. Music listening intervention will consist of 30 minutes of music listening time each time a participant experience pain at a self-assessed level of 3 on the Numeric Pain Rating Scale (or 1 level before they generally take pain medication). Participants will provide their own music and music listening devices. Participants will record music listening sessions, medication use, hospital and emergency room visits, and overall mood after music listening session in an electronic journal. Data from patients will be analyzed post- intervention.

Specific Aims and Data Analysis:

Aim 1 – Determine the social patterning of the effect of music listening. This aim will be accomplished by conducting multivariate linear regression analyses in SPSS to determine which characteristics are statistically related to a positive effect.

Aim 2 – Determine the association between music listening and chronic pain perception. This aim will be accomplished by conducting multivariate linear regression analyses in SPSS to determine the statistical association between music listening and chronic pain perception. An independent samples t-test will be used to determine if there are significant group mean differences in the variability of chronic pain perception between the participants in the experimental group and participants in the control group at the start and completion of the 4 week music therapy intervention. A paired samples t-test will be used to determine if there are individual mean differences for each participant at the start and completion of the 4 week music therapy intervention.

Aim 3 – Determine the association between music listening and hospital visits. This aim will be accomplished by conducting multivariate linear regression analyses in SPSS to determine the statistical association between music listening and hospital visits. An independent samples t-test in SPSS will be used to determine if there are significant group mean differences in the variability of hospitals visits between the participants in the experimental group and participants in the control group at the start and completion of the 4 week music therapy intervention. A paired samples t-test in SPSS will be used to determine if there are individual mean differences for each participant at the start and completion of the 4 week music therapy intervention.

Aim 4 – Determine the association between music listening and emergency room visits. This aim will be accomplished by conducting multivariate linear regression analyses in SPSS to determine the statistical association between music listening and emergency room visits. An independent samples t-test in SPSS will be used to determine if there are significant group mean differences in the variability of emergency room visits between the participants in the experimental group and participants in the control group at the start and completion of the 4 week music therapy intervention. A paired samples t-test in SPSS will be used to determine if there are individual mean differences for each participant at the start and completion of the 4 week music therapy intervention.

Aim 5 – Determine the association between music listening and medication usage. This aim will be accomplished by conducting multivariate linear regression analyses in SPSS to determine the statistical association between music listening and medication usage. An independent samples t-test in SPSS will be used to determine if there are significant group mean differences in the variability of medication use between the participants in the experimental group and participants in the control group at the start and completion of the 4 week music therapy intervention A paired samples t-test in SPSS will be used to determine if there are individual mean differences for each participant at the start and completion of the 4 week music therapy intervention.

REFERENCES

Akombo, D. (2009). *Music and Medicine: Connections Found*, New York, NY: Seaburn Press.

Ballas, SK (2007). Current issues in sickle cell pain and its management. *Hematology American Society of Hematology Education Program* 2007: 97-105.

Ghaffaripour, S, Mahmoudi, H, Sahmeddini, MA, Alipour, A, and Chohedri, A. (2007). Music can effectively reduce pain perception in women rather than men. *Pakistan Journal of Medical Science,* 29(1), 128-131.

Linnemann, A, Kappert, MB, Fischer, S, Doerr, JM, Strahler, J, and Nater, UM. (2015). The effects of music listening on pain and stress in the daily life of patients with fibromyalgia syndrome. *Frontiers in Human Neuroscience*, 9(434). doi: 10.3389/fnhum.2015.00434

Marlowe, KF and Chicella, MF. (2002). Treatment of Sickle Cell Pain. *Pharmacotherapy: The Journal of Human Pharmacology and Drug Therapy*, 22(4), 484-491.

McCaffery, M., Beebe, A., et al. (1989). *Pain: Clinical manual for nursing practice,* Mosby St. Louis, MO.

Mitchell, LA and MacDonald, RA. (2006). An experimental investigation of the effects of preferred and relaxing music listening on pain perception. *Journal of Music Therapy* 43(4), 295-316.

Mitchell, L. (2007). A survey investigation of the effects of music listening on chronic pain. *Psychology of Music*, 35, 37-57.

Pothoulaki M, Macdonald RA, Flowers P, Stamataki E, Filiopoulos V, Stamatiadis D, Stathakis CP. (2008). An Investigation of the effects of music on anxiety and pain perception in patients undergoing haemodialysis Treatment. *Journal of Health Psychology* 13, 912-920.

Thompson, W. (2013). Factors influencing the use of Complementary Alternative Methods (CAM) in patients attending an adult sickle cell clinic in Jackson, Mississippi. *Journal of Behavioral Health*, 2(2), 151-158.

Thompson, W. (2014). Pain Control in Sickle Cell Disease Patients: Use of Complementary and Alternative Medicine. *Pain Medicine*, 15(2), 241-246.

University Hospital Case Medical Center; Case Western University. Music therapy in sickle cell pain mixed methods study. In: ClinicalTrials.gov [Internet]. Bethesda (MD): National Library of Medicine (US). 2000- [cited 2014 October 29]. Available from: https://www.clinicaltrials.gov/ct2/show/NCT02270060 NLM Identifier: NCT02270060.

Carolyn Graham's Jazz Chants

Roy Kennedy

Abstract

The purpose of this paper was to write a descriptive analysis of Carolyn Graham's jazz chants and to provide evidence from music therapy studies, which is related to jazz chants. Graham created jazz chants as an English as-a-Second Language (ESL) teaching technique. I interviewed Ms. Graham via email communication, reviewed several of her jazz chants books and accompanying CDs, and summarized evidence from experimental and qualitative music therapy studies as sources for this paper. Two of the music therapy studies I reviewed provided evidence, which indicated that the pairing of second language information with music stimuli was effective in developing listening comprehension and expressive language skills in both Kindergarten and middle school ESL students. Another music therapy study showed that pairing language information with singing and sign language activities was valuable in assisting Kindergarten ESL students to acquire vocabulary words. Overall, the results of music therapy interventions used as supplementary and/or primary methods of teaching ESL skills to Kindergarten, elementary, and middle school students were shown to be effective ESL teaching techniques. The music therapy studies reviewed for this project used Graham's basic jazz chants philosophy of rehearsing language information through rhythmic repetition, the chunking of language information paired with melodies and chants, and modifying the affect of second language learners as useful strategies in ESL instruction.

I first became aware of jazz chants while attending the National Association of Music Therapy Conference during 1985 in New York City. As I walked down the hallway of the conference, I heard some vigorous chanting, singing, and finger snapping activities emanating from one of the concurrent sessions. The contagiously energetic and lively atmosphere generated from this presentation drew me into the room as the attendees were chanting and singing along with Ms. Graham while learning ESL content.

Ms. Carolyn Graham was born in 1931 and grew up in Benicia, California. Some of her favorite activities during childhood were playing the piano and learning to ride horses. She began to play the piano at the age of four, taking lessons from her father, a math teacher, who was a classical musician and also conducted a small orchestra. Graham had a classical piano teacher from Leipzig as well, but her dream was to play the music of her father, ragtime piano, and travel the world (personal communication, C. Graham, March 23, 2014).

Graham, however, was not that interested in becoming a teacher. As she recalled, "that sounded like a boring life. I just wanted to run around the world" (personal communication, C. Graham, March 23, 2014). In fact, she wasted no time in satisfying her yearning to travel as she graduated from The University of California, Berkeley.

At the age of 23, Graham immediately journeyed to Turkey, and she first taught English there when asked to fill in for a teacher that was ill.

She remembered, "I went into the class and lo and behold, I adored it!" (personal communication, C. Graham, March 23, 2014).

Upon returning to the U.S., Graham began teaching ESL classes by day and playing jazz in the piano bars of New York City at night. I was very involved with traditional American jazz, four beats to a bar. I loved playing in the piano bars in New York. At that time I was teaching English just to pay the rent but at night I was working in the bars, singing and playing the old jazz (personal communication, C. Graham, March 23, 2014).

During one of her moonlighting gigs in New York City, Graham had an epiphany that led to the creation of jazz chants. I had just finished a song when a friend of mine came up to the piano and said, "Gee, it's good to see you. You look wonderful!" I heard exactly the 4/4 rhythm of the music I was playing and at that moment I wondered if it were true that American English had the same sound system as traditional American Jazz. I started listening very carefully to everything people were saying and I heard that four beat everywhere. Jazz chants was born! (personal communication, C. Graham, April 4, 2014).

Graham published her first book, *Jazz Chants*, in 1978. She was a Master Teacher of ESL at New York University as part of the American Language Institute, a Teaching Fellow at Harvard University, and was a recipient of several U.S. State Department teaching grants. Additionally, Graham has taught ESL and trained teachers in South Africa, Egypt, West Africa, Ethiopia, Mali, Russia, Latvia, Lithuania, Georgia, the Ukraine, and Peru. From the Eastern Canadian Arctic to Madagascar, Graham has proliferated jazz chants around the world (Jazz Chants, 2014). When I asked how many people are currently practicing jazz chants, she responded with the following answer:

> I have given training workshops pretty much all over the world. I know the chants are widely used in Japan, China, Korea, and Malaysia. Also I spent some time in Africa and Madagascar and was very well received. I have had splendid audiences of teachers and students all over Europe and I continue to get emails from Afghanistan with pictures of young women smiling and holding my Jazz Chant books (personal communication, C. Graham, September 4, 2014).

The Definition of Jazz Chants

Graham defined jazz chants as, "the rhythmic expression of Standard American English as it occurs in situational contexts" (Graham, 1978, p. 9). She originally designed jazz chants as a language acquisition tool, which emphasized the rhythm and intonation patterns of spoken American English.

Graham emphasized the Development of "listening comprehension," which is crucial to the success of ESL students. The idea that English Language Learners (ELLs) must be able to hear and understand native speakers in order to first understand and then speak a second language is central to her teaching approach. Native speakers of English often distort the language by shortening, lengthening, or dropping the sounds of words, which makes normal conversation almost impossible to comprehend for ESL students unless their ears have been trained to hear such caveats (Graham, 2001).

How Jazz Chants Work

Most jazz chants are presented in two-part dialogue form as the teacher engages the class in call and response activities. First, the teacher models a line of the chant and the group responds in unison. This type of choral response takes place for each line of the chant and the students are encouraged to clap, play rhythm sticks, or snap their fingers in order to establish a clear and steady beat (Graham, 2001).

During the second step, the class is divided into two equal sections and each section of the class responds as alternating groups while the teacher models each line of the chant. Graham adds, "there is no limit to the number of students that may be involved in each section as jazz chants may be conducted with two students or two hundred students" (Graham, 1978, p. 13).

During the third step, the chant is conducted as a two-part dialogue; the teacher gives the first line of the chant and the class responds in unison with the second line of the chant. This technique continues until the class is comfortable with the contents of the chant. The groups then respond to each other in two-part dialogue without the teacher as a model. Each group simply responds to the teacher's hand cue as they chant in an antiphonal choir like manner (Graham, 1978).

In Graham's *Children's Jazz Chants* book, Graham describes the cloze procedure in which students fill in the blanks of lyric lines or chants with a missing word or phrase. She specifies that this activity can be performed either orally or in writing; Graham adds that method of performing the cloze procedure is to chant or sing part of a line, then stop, allowing the students to chant or sing the rest of the phrase. This technique is similar to the Musical Speech Stimulation (MUSTIM) music therapy intervention, which leaves blanks in the overlearned lyrics of songs, rhymes, chants, and musical phrases to stimulate speech in patients with apraxia or aphasia (Davis, Gfeller, Thaut, 2008).

Jazz chanting immerses ESL students in expressing feelings in situational contexts by emphasizing the stress and intonation of the language while simultaneously building vocabulary that is used in everyday conversations. The chants are designed to portray basic human emotions such as pleasure, joy, anxiety, feeling homesick, physical pain and misery, and the chants often use evocative or exclamatory statements to elicit responses from the students. For example, in one chant, the feelings of frustration, over misplacing an object is role-played as a game.

An individual student wanders around the classroom as if searching for a misplaced item, while asking the appropriate question, "Where is it?" The rest of the class offers helpful suggestions as group responses, and the emotion of "relief" is expressed at the end of the chant when the item is found (See Table 1).

Table 1

Excerpt from the jazz chant
It's Got to Be Somewhere

Where is it? Where is it?
        Where is it? Where is it?
I can't find it!
        She can't find it!
It's got to be here, it's got to be here!
        Take it easy! Take it easy!
Here it is! Here it is!
        Whew!

Note. The group responses are indicated with indented phrases. This excerpt was taken from Ms. Graham's book *Jazz Chants* (1978). Reprinted with permission from Ms. Carolyn Graham.

Music Therapy Studies that Use Jazz Chants Techniques

There is a small but growing body of music therapy research literature that generally supports the basic philosophy and tenets of Graham's jazz chants. One study in the music therapy literature (Kennedy & Scott, 2005) supports Graham's claims that ESL students' language acquisition may be enhanced when they are involved in pairing second language information with music stimuli. In this study, the English speaking skills ($p<.01$) and story retelling skills ($p<.001$) of 34 middle school students were significantly increased from pretest to posttest as a result of their participation in active music listening, unison chanting while playing instruments, singing, rhythmic training, and lyric analysis and rewrite activities. The procedure of this study involved high and low-functioning 6th and 7th grade ESL students using Kodaly rhythm syllables that were paired with target vocabulary words and phrases of lyrics in songs while the syllables and lyrics were chanted and played on hand drums. Additionally, the cloze procedure was used as students identified key words in song lyrics left as blanks on a dry erase board after hearing and singing complete verses of the songs. The therapists rewrote verses of songs on the dry erase board and engaged the students in lyric discussions paired with pictures that supported topical conversations and story retelling skills as well.

Related to the idea of developing listening comprehension, a descriptive music therapy study (Kennedy, 2008) indicated that pairing language information with listening, chanting, group singing, playing simple percussion instruments, and movement to music activities enhanced the English speaking and story retelling skills of Kindergarten ESL students. A community based, after-school ESL music therapy group scored 45.9% in the "superior" category over all of the skills on the Story Retelling Inventory as compared to the public school ESL group which scored 0.00% in the "superior" category over all the items. In addition, the after-school ESL music therapy group scored 55% in the "always" category in comparison to the public school ESL group, which scored 30% in the "always" category on the English Speaking Checklist. The Kindergarten ESL students responded enthusiastically to chant activities that required them to identify the types of food they liked to eat using the fill in the blank cloze procedure. Students played rhythm sticks during group chants, which allowed the therapist to create chant dialogues with the students. For example, "what do you do when you're feeling hungry?" Answer: "Eat food" (Kennedy, 2008, p. 99).

Group singing was also useful during this study in order for the students to practice specific phonemes such as the "pl" and "th" sounds. The therapists slowed the tempo of the songs and enunciated the targeted phonemes with great emphasis in efforts to assist the students in acquiring certain word sounds in the context of song lyrics.

Related to the idea of building vocabulary and expressing feelings in situational contexts, Schunk (1999) conducted a study that investigated the effects of singing paired with signing on the receptive vocabulary skills of elementary ESL students. This study involved eighty children (K-2) in one of four rehearsal conditions: sung text paired with signs, spoken text paired with signs, sung text, and a spoken text only control group.

The results of this study indicated that the sung text paired with signs and the spoken text paired with signs conditions yielded significantly greater acquisitions of vocabulary recognition ($p < .05$) than the spoken text only control condition.

Schunk stated that, "music has become a valued tool in theoretical design and practical application of English-as-Second Language instruction" (Schunk, 1999, p. 111). She indicated that Krashen (as cited in Schunk, 1999) distinguished between language "acquisition," which is a subconscious internalization of language information, and language "learning," which requires conscious effort as in the memorization of grammar rules. Schunk furthered this distinction between language acquisition and learning by citing Krashen's "Din in the Head" phenomenon, in which second language students involuntarily rehearse language information in their heads in much the same way as musicians cognitively rehearse music without playing instruments or singing.

Guerrerro (as cited in Schunk, 1999) furthered this explanation by indicating that ESL students at all levels of English proficiency engage in subconscious Din rehearsal, which has positive practical and emotional effects on the language student, including additional practice in the language and greater confidence in speaking the language (pg. 111).

Murphey (as cited in Schunk, 1999) further elaborated on the Din theory by suggesting that songs may act as language acquisition devices. He claimed that "having a song or part of a song running through one's mind is equivalent to hearing inner speech, also known as Din" (pg. 112). Murphey called this concept the song-stuck-in-my-head (SSIMH) phenomenon and collected data from 49 participants on a questionnaire in which all of the participants reported having the SSIMH experience. Furthermore, second language instruction research exists which reports that specific language goals may be achieved with the controlled application of songs.

Songs can effectively reinforce teaching by helping to practice and revise vocabulary, idioms, sentence patterns, pronunciation, stress, rhythm, and intonation in a variety of language styles, and offer background cultural information without recourse to barren drill (Rees, as cited in Schunk, 1999, p.112).

Research in music therapy literature supports the idea that language may be learned through the use of song lyrics, especially when the lyrics are paired with "good Gestalts---well-organized rhythmic, melodic and harmonic patterns, which can enhance the perceptual process, as we filter, select, and organize information for further processing" (Davis, Gfeller, Thaut, 2008, p. 59). For example, interesting or preferred music may act as a mnemonic device, which assists in the encoding, processing, and storage of the language information in the long-term memory. The rehearsal of language information through repeated song lyrics allows new information to be linked with familiar melodies, which may enhance the learning of language information, e.g. the letters of the alphabet set to the song Twinkle, Twinkle, Little Star.

Related to the idea of pairing language information with music and other extra-linguistic stimuli, Lim (2012) was successful in the use of singing paired with pictures of target words and phrases in the development of speech-language skills for children with Autism Spectrum Disorder (ASD). Lim composed six songs in which she left a blank at the end of each line of the song for a target word or two-word phrase. As Lim sang the words to each song, a picture representing the target word or phrase was held up for the student to see so he/she could associate the target word or phrase with a visual stimulus.

The results of Lim's study indicated that children with autism significantly improved their verbal production ($p < .05$) in both music training and speech training groups in comparison to a no-treatment control group. However, the music group attained higher scores on the verbal production scale than the speech group, and the children assessed as low-functioning were able to acquire more target words and phrases in music training than in speech training. She also noted that children assessed as low-functioning responded with more attention and understood much more of the language information as it was paired with recognizable patterns in the music (Lim, 2012).

The technique of pairing language information with extra-linguistic aides in the form of audio and visual stimuli is a highly recommended strategy in regular ESL instruction. Krashen (2003) argued that comprehensible input is the key to language acquisition and that the use of comic books, audiotapes, and videotapes, enhances the students' acquisition of a second language. Graham has also supported the idea of using audio and visual aides in ESL instruction. Examples of Graham's jazz chants and a movie that she created are available to the public free of charge; The jazz chants may be downloaded in MP3 format (jazzchants.net) along with the movie, *Fish Dreams*, in FLV format.

Music Therapy, ESL, and Cultural Considerations

Another music therapy study (Schwantes, 2009) described the use of music for the purpose of enhancing the ESL skills of 6 Latino children in a public school. Four of the children were Kindergarten students that participated in half-hour group sessions using traditional Spanish language children's music which had been translated into English, and two additional students, a fourth grader and a sixth grader, attended music therapy ESL sessions twice weekly and used music from Western pop culture. The regular ESL teacher referred the students to music therapy for a total of 12 sessions. During the first three sessions, the music therapist observed the students in their regular academic classrooms during group instruction while collecting assessment data concerning the students' response to questions, length of utterances, accurate comprehension, and their ability to sequence events from a song or sound story. The music therapist then designed and implemented movement to music, instrument playing, singing, and verbal fill-in-the-blank interventions as supplemental ESL instruction. Students were also asked the "W" questions (e.g., who, what, when, where, why) after singing songs to assess their understanding of the contents and sequence of the songs.

All of the Kindergarten students and both the elementary and middle school students showed improvement in increased vocabulary, receptive and expressive language skills, and sequencing skills as well. Perhaps the most important finding of this study was that using a combination of music from the students' culture and the culture in which they live can be effective in designing music therapy interventions for ESL students.

In addition, another music therapy study (Stock, 2009) reported that bilingual music therapy interventions were of benefit to ESL students that had special needs. Eleven ELLs participated in an English-only music therapy control group in comparison to a bilingual music therapy experimental group. In the English-only group the researcher used Western songs and spoke and sang only in English, while in the bilingual group the researcher sang folk songs and spoke with the students in both English and Spanish. There was no significant improvement on the Woodcock-Muñoz Language Survey-Revised test (WMLS-R). More improvement was documented, however, in expressive language skills on the Pre-Kindergarten Assessment Chart (PKAC) for the bilingual group than the English only group.

Applications of Music Therapy Using Jazz Chants Techniques

There is a need for effective ESL instruction as the numbers of students with Limited English Proficiency (LEP) continues to grow. The Government Accountability Office (as cited in Adamek & Darrow, 2010) identified approximately 10% of the total student population in the U.S. as having LEP in 2006. More recently, the Associated Press identified children speaking a language other than English at home as one of the fastest growing populations in the U.S. Their numbers doubled between 1980 and 2009, and now they make up 21 percent of school-age kids (Armario, 2013, para. 3). Students with LEP have difficulty in speaking, reading, writing or understanding the English language. This deficit severely impedes their progress in learning academic subject matter in schools where English is the only language of instruction (Adamek & Darrow, 2010; Armario, 2013). Students with LEP have some of the most significant challenges in academic achievement of all the minority groups in the public schools and have been identified as an "at risk" group with high dropout rates. Recent evidence of their academic difficulties was documented on a nationwide reading exam; only 7 percent of fourth-grade and 3 percent of eighth-grade ELL students achieved proficiency, while many ELL students remained stymied in ineffective ESL programs (Armario, 2013).

"Students with limited English, whether receiving LEP services, special education and related services, or a combination, will be a part of the regular education inclusive music classroom, adapted music classroom, or music therapy services" (Adamek & Darrow, 2010. p. 8). Since music therapists already provide interventions in the form of direct and consultant services for students that have developmental, cognitive, and social skills challenges in special education and music education settings, it seems logical that music therapy services may be included on the Individualized Education Plans (IEP) of students with LEP as well.

In fact, there already exists some evidence in music therapy research literature, which indicates that music therapy interventions may be effective ESL instruction in the following scenarios: as a supplementary teaching technique in regular ESL classrooms, as additional ESL instruction in pull-out classrooms, and as primary ESL instruction in community-based after-school tutoring programs.

Graham's jazz chants are creative, imaginative, colorful, and evocative music/language compositions that elicit motivated and enthusiastic responses from ESL students of all ages. The effectiveness of jazz chants is based on the concept that pairing language information with engaging rhythmic and melodic music stimuli assists students in the acquisition, rehearsal, and speaking of a second language. In any case, it seems that Graham's jazz chants may be a valuable resource for music therapists in creating stimulating and engaging music therapy interventions for ESL students in several different settings.

References

Adamek, M.S. & Darrow, A.A. (2010). *Music in special education* (2nd ed.). Silver Spring, MD: The American Music Therapy Association.

Armario, C. (2013, April 14). English-language learners face shortage of teachers, and successful bilingual programs. *Associated Press*. Retrieved from http://nbclatino.com/2013/04/14/english-language-learners-face-shortage-of-teachers-and-successful-bilingual-programs/

Davis, W.B., Gfeller, K.E., & Thaut, M.H. (2008). *An introduction to music therapy: Theory and practice* (3rd ed.). Silver Spring, MD: The American Music Therapy Association.

Government Accountability Office. (2006). *No Child Left Behind Act: Assistance from education could help states better measure progress of students with limited English proficiency*. Report GAO-06-815 (July). Washington, DC.

Graham, C. (1978). *Jazz chants*. New York, NY: Oxford University Press.

Graham, (2002). *Children's Jazz Chants: Old and new* New York, NY: Oxford University Press.

Graham, C. (2001). *Jazz chants: Old and New*. New York, NY: Oxford University Press.

Kennedy, R. (2008). Music therapy as a supplemental teaching strategy for Kindergarten ESL students. *Music Therapy Perspectives, 26*(2), 97-101.

Kennedy, R. & Scott, A. (2005). A pilot study: The effects of music therapy interventions on middle school students' ESL skills. *Journal of Music Therapy, 42*(4), 244-261.

Krashen, S. D. (1983). *The Natural Approach*. New York: Permagon Press.

Krashen, S. D. (2003). Explorations in language acquisition and use. Portsmouth, NH: Heinemann.

Jazz Chants (2014). *Who is Carolyn Graham*? Retrieved from jazzchants.net.

Lim, H. (2012). *Speech and language development through music for children with autism*. Philadelphia, PA: Jessica Kingsley.

Schunk, H. A. (1999), The effect of singing paired with signing on receptive vocabulary skills of elementary ESL students. *Journal of Music Therapy, 36*, 110-124.

Schwantes, M. (2009). The use of music therapy with children who speak English as a second language: An exploratory study. *Music Therapy Perspectives, 27*, 80-87.

Stock, C. (2009). *The effect of bilingual music therapy on the expressive language of Spanish-speaking children learning English in early childhood classrooms*. (Unpublished master's thesis). Texas Woman's University, Denton, Texas, U.S.A.

Bio-guided Music Therapy and the Clinician/Researcher Model

Eric B. Miller, Ph.D. MT-BC
Montclair State University
David Ott Lab for Music & Health

While alternative medicine is becoming more accepted and integrated into mainstream healthcare, there is a concurrent demand by insurance companies and program reviewers to provide efficacy analyses and data supporting best practices. Clinical research trials as well as individual case research help inform the overall knowledge base related to practice techniques and outcomes. The dual-missioned Clinician/Researcher, is "an individual who both provides behavioral health services and conducts research involving human participants that is related to the provision of such services" (Yanos, P. T., & Ziedonis, D. M., 2006). There are advantages and disadvantages to taking this dual-tasked position, as well as potential ethical considerations.

Bio-guided Music Therapy (Miller, 2011) provides optimal opportunities for the integration of clinical music therapy practice with music therapy research. Employing a baseline - intervention design, the clinician establishes an initial set of physiological measurements by which to compare subsequent music therapy or adjunctive interventions. An example would be taking readings from a finger sensor that provide readings of electrodermal response as an indicator of sympathetic nervous system arousal, or from a photoplethysmograph to take heart rate and blood volume pulse readings. Alternatively, neurological measures could be taken, such as EEG brainwave rhythms. Following baseline readings, physiological measurements continue to be taken during music therapy interventions which may include music listening, active improvisational music making, lyric analysis, etc. Stress indicators, such as electrodermal activity and heart rate, are quick to show increases in arousal while listening to particular sections or selections of music, and may spike when discussing lyrics that of poignant significance for a patient/subject. These elevated readings are vital informational clues for the music therapist that may help guide music intervention in real time while the intervention is unfolding.

Advantages to having such physiological data at your fingertips, is the ability to quickly assess where progress appears most accelerated and which music therapy interventions were associated with that improvement. Replication of the intervention may then help to determine if the effect is reproducible or if it occurred by chance, or some unexplained variable. A graphical representation of the data from a session is useful for easy illustration of the effects of the music therapy to both clients and other professionals, including doctors, nurses and insurance evaluators. Statistical summaries may also be useful for evaluating efficacy and demonstrating.

Bio-guided music therapy offers a solution to the perceived disparity or "clash" in the dual role position. According to Yanos and Ziedonis, the clinician-researcher may experience an internal clash between the clinical mandate to act in the patient's best therapeutic interest and the scientific mandate to pursue truth with all appropriate rigor (Yanos, P. T., & Ziedonis, D. M., 2006, p.3). In Bio-guided music therapy, this conflict is reduced due to the integration of physiological goals into the outcome objectives for the patient. Through this process, research data may be acquired concurrently and seamlessly with the patient's clinical therapy.

References

Miller, E. B. (2011). *Bio-Guided Music Therapy: A practitioner's guide to the clinical integration of music and biofeedback*. Jessica Kingsley Publishers.

Yanos, P. T., & Ziedonis, D. M. (2006). The Patient-Oriented Clinician-Researcher: Advantages and Challenges of Being a Double Agent. *Psychiatric Services (Washington, D.C.)*, *57*(2), 249–253. http://doi.org/10.1176/appi.ps.57.2.249

The Effect of Preferred Listening on the Anxiety, Nausea, and Claustrophobia of Cancer Patients During Radiation Treatments

Pomi Yun BS, MPH & Roy Kennedy PhD
University of Georgia

Abstract

Purpose: The purpose of this study was to investigate the effects of cancer patients listening to their preferred music on claustrophobia, anxiety, and nausea during radiation treatment. Patients with head and/or neck cancer may experience claustrophobia as a result of having to wear a mask during their radiation treatments and; in general, many patients with cancer below the head have anxiety and nausea as a result of anticipating and receiving radiation treatments as well.

Method: This quasi-experimental pilot study included a single group of volunteer participants that were exposed to music during their treatments. The participants were asked to indicate their preferred artists and songs, which were played via Pandora on an iPod during their radiation treatments. Initially, customized playlists were created for the patients; however, the patients quickly became exhausted of the music and asked for more variety. Through artist-preferred playlist provided by Pandora, the patients were more satisfied. Data was collected via paper and pencil questionnaires for the dependent variables anxiety, claustrophobia, and nausea. The researchers and nurses sat with patients prior to and after their first radiation treatment and final radiation treatment to assist the patients with any questions about the questionnaires. Only patients that could not cognitively understand the questions after the researchers read the questions aloud and gave repeated prompts were excluded from participating in the study. Since the treatment room was very small, the patients' preferred music was played on an Ipod through connected speakers. The radiation therapist controlled the music source during each patient's treatments. The researchers calculated the scores from the questionnaires and compared the pretest to posttest mean gain scores from the first treatment to the pretest/posttest mean gain scores for the last treatment.

Results: Seven out of thirteen patients (53.8%) showed a decrease in suffocation related to claustrophobia. Ten out of thirteen patients (76.9%) showed a decrease in restriction related to claustrophobia. Eight out of thirteen patients (61.5%) showed a decrease in levels of anxiety. The patients' scores on the nausea variable showed no changes or was never present throughout the treatment period. The researchers recorded qualitative comments that the patients made before and after treatments, which provided additional information concerning their experiencing of the dependent variables.

Conclusion: The results of this study may be of use to music therapy professionals that work with cancer patients undergoing radiation therapy. The majority of music therapy studies with cancer patients has focused on those undergoing chemotherapy treatment. In addition, no music therapy studies were found that specifically investigated the effects of patient preferred listening on the claustrophobia of cancer patients.

What is Radiation Therapy?

Cancer diagnosis is where the stress begins for a patient. The stress and anxiety continue to escalate as the patient receives treatment options and plans (McGregor & Antoni, 2009). Radiation therapy (RT) is often used for cancers of the breast, prostate, lung, and head and neck area. This curative therapy targets areas using radiation to attack the rapidly growing and dividing cancer cells. Unlike chemotherapy, setting dosage limitations and spreading the RT over a period of three to seven weeks can protect the normal cells. After consultation with an oncologist, a patient is seen at a simulation session. At the simulation session, the treatment plan for the patient is created and finalized. In this particular study, the patients were not seen at the simulation session to allow patients to adjust to the first exposure to the devices and machines that can cause a patient anxiety (Clark, 2006).

Patients specifically receiving radiation therapy of the head and/or neck area experience a unique unsettling pain and negative effects along with the stress. This is due to the treatment protocol that requires a patient to lie completely still while wearing a molded, netted-mask that is secured to the table on which the patient lies. This experience can produce pain in sensitive areas such as the throat for patients receiving radiation in that specific area which may last until their last treatment (treatments lasting up to six weeks or longer). Patients typically receive radiation therapy five days consecutively per week for three to seven weeks on the same area that is continuously irradiated (Clark, 2006).

The treatment area can often become extremely sensitive thus producing pain and often irritation, which can be temporarily remedied with medical creams. These extensive treatment schedules may last longer due to unforeseen events such as hospitalizations for non-cancer related medical issues such as pneumonia or simply due to patient absences. For example, many patients may simply skip and/or avoid treatments as the side effects of the radiation treatments escalate. Just as chemotherapy has shown significant negative impacts on patients' (QOL), radiation may also induce similar stresses and burdens of disease (Vargo, Ferris, Clump, & Heron, 2014). The high emotional stress can lead to depression and anxiety that may decrease a patient's overall quality of life (QOL). Patients that experience such emotional and psychological stressors become less compliant with radiation treatments and more difficult to treat (Burns, 2001).

Health Benefits of Music Therapy

The effects of music therapy treatment for cancer patients have resulted in milestone improvements in areas such as depression, anxiety, and stress. Mindful music listening has shown improvement in patients who suffer from depression (Eckhardt & Dinsmore, 2012). Not only have patients with depression experienced decreased levels of depression with repeated exposures to listening to preferred music but their levels of stress and anxiety have also been significantly decreased when compared to a control group that received no music (Mohammadi, Shahabi, & Panah 2011).

Music therapy has demonstrated its powerful abilities to transcend its positive effects in addressing medical conditions and also improving overall mood, which can contribute to treatment outcomes (Jamabo & George, 2014). A study in older adults has also shown a reduction in levels of depression, which shows that music therapy transcends age and can be applicable to many different age groups facing similar medical situations (Chan et al., 2012). Music therapy can benefit patients by improving overall health outcomes in areas such as surgery, cardiology, and oncology (Aldridge, 1993). Studies that focused on factors such as preferred music selection and familiarity have shown greater decreases in levels of stress than studies that used non-preferred music (Pelletier, 2004). Music therapy has provided an enhanced quality of life for many patients that have incurable or difficult to treat diseases.

Music Therapy and Radiation Treatment

Reports have shown that music therapy in cancer treatment assists in alleviating pain perception along with decreasing nausea (Curtis 1986; Sabo & Michael, 1996; Standley, 1992; Zimmerman et al., 1989). Furthermore, radiation therapy has been studied in the past to measure the specific effect of music therapy on the level of anxiety in patients (Smith, Casey, Johnson,

Swede, & Riggin, 2001). In the study of 42 male subjects with a variety of cancer types, the treatment group listened to preferred music and showed a statistically significant decrease in the state anxiety over time. Overall, there was no distinct difference on overall performance between the treatment and control groups.

This study specifically attempted to measure the changes in levels of claustrophobia during curative RT by adopting a similar approach to previous music therapy studies involving RT but focused on quantifying the levels of claustrophobia while similarly measuring anxiety and nausea. The purpose of this study was to observe the effects of preferred listening on cancer patients while specifically measuring claustrophobia, anxiety, and nausea during the radiation treatments.

Method

Participants

The volunteers were recruited from a radiation treatment facility in a small town in the southeastern part of the United States. The participants were recruited according to their willingness to participate in the study. They were informed about the requirements of their participation at their consultation before the first radiation treatment began. Upon agreeing to participate in the study, the participants had a simulation in order to customize their treatments. The study did not start at the simulation, however, because the researchers did not want to overwhelm the patients at the first exposure to the large magnetic machines and masks. Also, the simulations took place in a different room than where normal treatment took place. Upon completing simulation the patients were then scheduled for their first day of radiation, which was the first day that they received the pre-treatment questionnaire and then post-treatment questionnaire immediately following RT.

From their first treatment, music of the patients' preference was played during each of their treatments five days a week for however many weeks they were prescribed RT. On the last day of their treatment, the participants again filled out the questionnaires before and after treatment. All of the participants signed informed consent forms prior to the start of the study.

Materials

An iPod was connected to a speaker that was placed in the radiation chamber using the Pandora application to play the patient's preferred music choices. Wi-Fi was available in the treatment room, which made Pandora a viable option.

A demographics questionnaire and three additional questionnaires were used as measurement tools for the following: claustrophobia, anxiety, and nausea. The demographics questionnaire was given only once to the participants before their first radiation treatment, which provided information about their age, gender, socioeconomic status, and music preference.

The Claustrophobia Questionnaire (CLQ) is designed to measure distinct symptoms that are hypothesized to make up claustrophobia: fear of suffocation and fear of restriction. For example, having a hard time breathing through your nose when having a cold accounts for fear of suffocation while having a difficult time sitting in the center of a crowded area accounts for fear of restriction. The participants were asked to rate a list of 26 items from a scale of 0 to 4 (0 = no anxiety, 4 = very anxious) in the given situations. Specifically the Y-1 Form of the State-Trait Anxiety Inventory (STAI) was administered to the entire group of participants before and after each of their radiation treatments. This test is designed to assess participant's state and trait anxiety with two parts. Each part contains 20 items with a 4-point rating scale (1 being not at all anxious and 4 being very anxious). The scores were calculated using the score key, translating raw scores into anxiety scores. A baseline score of 20 represented no presence of anxiety, while score higher than 20 represented levels of anxiety and scores lower than 20 indicated low anxiety or calmness.

The Nausea Questionnaire was also administered before and after the first and last radiation treatments. This test is designed to assess the subjective self-report level of nausea that the patient felt at the moment the questionnaire was administered. The participants were asked to mark on the visual analogue scale (VAS) line at the point they felt indicated the overall intensity of their nausea. The length of the VAS line is then recorded in centimeters. A long measurement of the VAS line indicated a person feeling very nauseous while a short measurement of the VAS line represented a person feeling hardly nauseous or no nausea.

Results

The median age of participants was 63.6 years with nine males and four females. All but one patient showed signs of the suffocation component of claustrophobia while all patients showed signs of the restriction component (Fig. 1 and Fig. 2). Seven out of thirteen patients (53.8%) showed a decrease in suffocation related to claustrophobia (Fig. 3).

Ten out of thirteen patients (76.9%) showed a decrease in restriction related to claustrophobia (Fig. 4). The baseline score for experiencing no anxiety was twenty where Patient K did not experience any anxiety before treatment or after the treatment period. Eight out of thirteen patients (61.5%) showed a decrease in levels of anxiety (Fig. 5). The patients' nausea scores showed no changes or was never present throughout the treatment period. The researchers recorded qualitative comments that the patients made before and after treatments, which provided additional information concerning their experiencing of the dependent variables. The patients commented on how they enjoyed listening to music and how they were able to hear new artists of genres they chose through the Pandora application. Overall, using Pandora versus the customized playlist resulted in positive feedback from the patients.

Figure 1. Suffocation Subscale

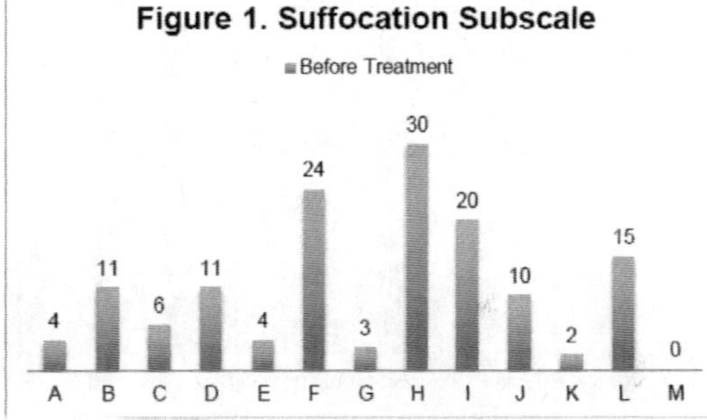

Figure 2. Restriction Subscale

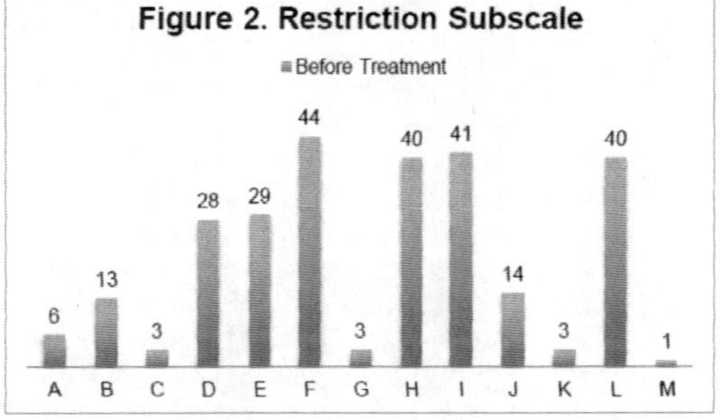

Figure 3. Fear of Suffocation

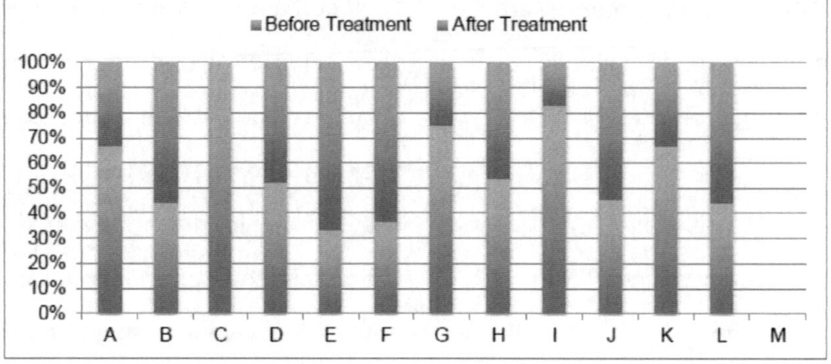

Figure 4. Fear of Restriction

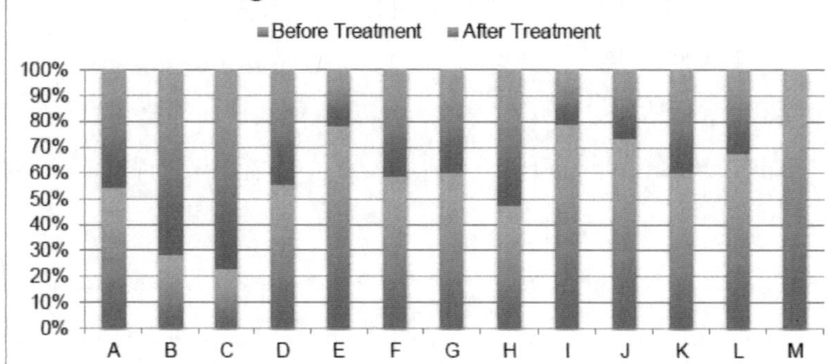

Figure 5. Anxiety

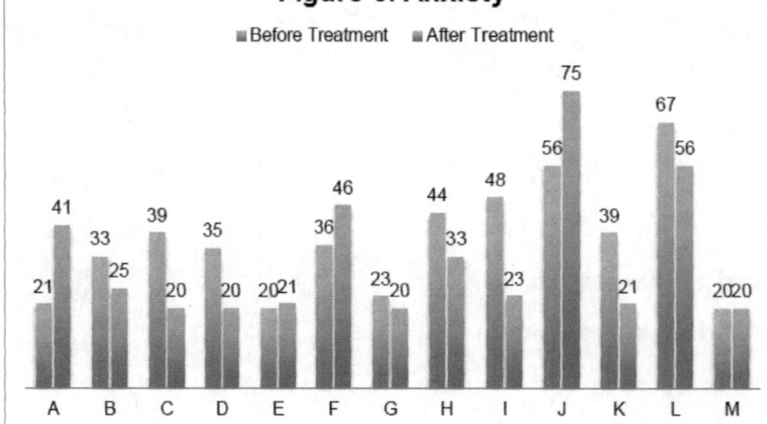

Discussion

The implications of music listening for patients undergoing RT for healthcare providers and treatment facilities are foundational. The results of this study indicate that the integration of music therapy into radiation treatment centers aids in creating a more comfortable environment for the patients. Creating this environment may in turn lead to more significant treatment results due to the less stressful impact of the treatment environment alone. Music therapy may also benefit the employees of treatment centers who endure every-day stressors from working with patients receiving radiation treatment.

Despite the benefits that have been demonstrated through improvements in health outcomes and increased relaxation through music therapy, it has still not been successfully integrated into all healthcare facilities (Burns et al., 2005).

A policy implementation could aid in the integration of music therapy in all oncology treatment facilities.

Limitations

Some of the limitations of this study include the following: There was a lack of experimental research literature concerning the use of music therapy in radiation treatment settings to support or corroborate the findings of this study. Furthermore, the sample size may not be representative of the larger population of patients that receive radiation treatments due to the selection of research participants from a local radiation treatment center in a small rural town. In addition, due to the longevity of radiation treatments that lasts six weeks on average, there were patients who dropped out of the study due to other medical complications, which resulted in a smaller sample size. Lastly, the results of the nausea assessments may not have indicated significant results because it is a more common symptom in chemotherapy rather than RT. Therefore, patients may have experienced nausea during chemotherapy treatment before arriving for radiation treatment. In addition, nausea could have been a symptom for patients receiving RT due to certain medications.

References

Aldridge, D. (1993). Music therapy research 1: A review of the medical literature within a general context of music therapy research. *The Arts in Psychotherapy*, *20*, 11-35.

Burns, D.S. (2001). The effect of the bonny method of guided imagery and music on the mood and life quality of cancer patients. *Journal of Music Therapy,* 51-65.

Burns, D.S., Sledge, R.B., Fuller, L.A., Daggy, J.K., Monahan, P.O. (2005). Cancer patients' interest and preferences for music therapy. *Journal of Music Therapy,* 185-199.

Chan, M.F., et al. (2012). Effects Of Music On Depression In Older People: A Randomised Controlled Trial. *Journal Of Clinical Nursing* 21(6),776-783.

Clark, M., Isaacks-Downton, G., Wells, N., Redlin-Frazier, S., Eck, C., Hepworth, J.T., & Chakravarthy, B. (2006). Use of preferred music to reduce emotional distress and symptom activity during radiation therapy. *Journal of Music Therapy, 43(3),* 247-265.

Curtis, S. (1986). The effect of music on pain relief and relaxation of the terminally ill. *Journal of Music Therapy, 23,* 10-24.

Eckhardt, K.J., Dinsmore J.A. (2012). Mindful Music Listening As A Potential Treatment For Depression. *Journal Of Creativity In Mental Health* 7(2),175-186.

Jamabo, T.A., George, I.R. (2014). Music As A Therapy For Depression In Women: A Theoretical Perspective. *International Journal Of Academic Research* 6(4),227-231.

McGregor, B.A., Antoni, M.H. (2009). Psychological intervention and health outcomes among women treated for breast cancer: a review of stress pathways and biological mediators. *Brain Behav Immun*, *23*(2), 159-166.

Mohammadi, A.Z., Shahabi, T. and Panah, F.M. (2011). An Evaluation Of The Effect Of Group Music Therapy On Stress, Anxiety And Depression Levels In Nursing Home Residents. *Canadian Journal Of Music Therapy* 17(1), 55-68.

Pelletler, C.L. (2004). The effect of music on decreasing arousal due to stress: A meta-analysis. *Journal of Music Therapy, 41,* 192-214.

Sabo, C., & Michael, S. (1996). The influence of personal message with music on anxiety and side effects associated with chemotherapy. *Cancer Nursing, 19,* 283-289.

Smith, M., Casey, L., Johnson, D., Swede, C., & Riggin, O.Z. (2001). Music as a therapeutic intervention for anxiety in patients receiving radiation therapy. *Oncology Nursing Forum, 28(5),* 855-862.

Standley, J. (1992). Clinical applications of music and chemotherapy: The effects on nausea and emesis. *Music Therapy Perspectives, 10,* 27-35.

Vargo, J.A., Ferris, R.L., Clump, D.A., Heron, D.E. (2014). Stereotactic body radiotherapy as primary treatment for elderly patients with medically-inoperable head-and-neck cancer. *Frontiers in Oncology, 4,* 1-20.

Zimmerman, L., Pozehl, B., Duncan, K., & Schmitz, R. (1989). Effects of music in patients who had chronic cancer pain. *Western Journal of Nursing Research, 11,* 298-309.

The Effect of Music on Impulsivity in College Undergraduate Students with Attention Deficits

Laura L. Dunbar

Abstract

The purpose of this study was to investigate the effect of music listening on impulsivity as judged by the Conners' Continuous Performance Test (CPT) II v. 5. College undergraduate students were recruited into one of two groups: one group of participants had no diagnosed history of ADHD while the other participant group had a history of ADHD. Both groups were administered a Continuous Performance Test (CPT) in an initial condition, a music condition, and a silence condition. All participants were tested in the initial condition with the remaining two (music and silence) randomly assigned to control for order effect. A significant main effect difference was found by group; the typical group exhibited lower impulsivity levels as compared to the ADHD group. Additionally, significant main effect differences were found by condition (initial, music, and silence) with all participants having lower impulsivity levels in the initial condition. No interactions by group among conditions. Implications and suggestions for future research were discussed. This document is a reduction of my dissertation; the full document is available on Proquest.

In the modern Western world, where music is readily available to everyone through radio, recordings, TV and videos and where recorded background music is routinely played in many public places, the need to understand the effects of music on our behaviour and cognitive processing has become increasingly important. (Hallam, Price, & Katsarou, 2002, p. 111)

People frequently listen to music while completing a variety of activities; however, the potential effects of background music on task completion is unknown. Research has been conducted investigating possible benefits of background music use, but the number of studies conducted regarding the effects of music on individual activity remains sparse (Kämpfe, Sedlmeier, & Renkewitz, 2011). More systematic investigation is needed to investigate potential effects of background music on learning among individuals with varied characteristics, specifically those with attention deficit disorders. The preliminary research requires further exploration to determine the consistency of such findings under controlled, experimental study conditions.

Background Music Use

Background music is frequently used in a variety of situations by the general population. Patton, Stinard, and Routh (1983) found 24% of surveyed elementary students and 40% of junior high students reported listening to music regularly while completing math homework (p. 284), with approximately 58% of all students indicating typical homework sessions included music listening (p. 285). Kotsopoulou and Hallam (2010) reported students selectively used music to concentrate while studying depending upon students' mood and subject studied; students used music more while writing and thinking than during test preparation, memorization, or foreign language study (p. 436). These data were not collected under controlled conditions, and while important, the data are based in self-perception.

Several researchers have studied the effect of music use on reading task achievement. Freeburne and Fleischer (1952) found no significant difference of music's effect on reading rates and comprehension among five different sound conditions—silence, classical music, semi-classical music, popular music, and jazz music—with the exception of a faster reading rate during the playing of jazz music. Henderson, Crews, and Barlow (1945) reported participants answering more questions incorrectly after reading a paragraph while listening to popular music as compared to both a classical music and a silence condition while testing reading efficiency. The authors suggested music complexity and task complexity may influence distraction levels (p. 317). Instrumental versions of show tunes (Fogelson, 1973) and hit songs as listed by Billboard Magazine (Anderson & Fuller, 2010) were found to be significant distractors on reading comprehension tests, while Hall (1952), found reading comprehension scores increased for 58% of participants in the presence of background music. Accuracy dropped during the no music conditions. Although mixed results were reported when examining music's effect on reading tasks, other tasks completed with background music have shown potential participant benefits. Background music has been shown to positively influence mood and states of arousal. Lesiuk (2010) found a significant increase in positive mood and increased cognitive performance when computer information system designers listened to individually preferred music while on the job. Music preference was found to be an important factor in stress reduction by both Lesiuk and by Jiang, Zhou, Rickson, and Jiang (2013), who found sedative music less relaxing than preferred music. Primary school students, ages 10 to 12, completed significantly more math problems and were able to remember more words in a memory task when "calming" music was played in the background (Hallam, Price, and Katsarou, 2002). Stanton's (1973) found similar results while investigating the benefits of playing classical music to high school and university students with high levels of anxiety, with students having high anxiety scoring significantly higher in the music condition as compared to a silence condition. Schellenberg, Nakata, Hunter, and Tamoto (2007) found various types of music increased arousal and mood while affecting cognition and cognitive testing results. These studies show impact of music on the general population; however, it is important to consider how arousal and stimulation differ between people with and without ADHD.

The Role of Arousal

"Given that music can directly impact arousal, with optimal levels of arousal increasing performance on certain cognitive tasks" (Robb, 2003, p. 278), arousal and stimulation theories could become a potential confound when studying students with attention deficits. Although other theories may also apply, three theories were chosen due to mention in the literature on which this research is focused: the Yerkes-Dodson Law of Arousal, the Optimal Stimulation Theory, and the Moderate Brain Arousal model.

Yerkes and Dodson (1908) noted habit formation was accelerated or maximized in a condition with moderate stimuli in mice (p. 481), which has been generalized to the human population. Humans typically exhibit higher arousal during sudden or surprising situations and lower arousal during more calm or monotonous conditions (Mehrabian, 1977). However, the Yerkes-Dodson Law of Arousal may not be sufficient to describe arousal patterns of those diagnosed with ADHD. Ornitz et al. (1997) found that boys with ADHD had an excessive pleasurable reaction to negative stimuli—loud, startling noises— in comparison to their typical male peers. Those with ADHD did not show a "sustained increase" in arousal as judged by heart rate, while their typical peers showed an increase in arousal as judged by heart rate throughout the experiment (p. 1704). Hence, potential exists for different responses to a variety of stimuli when researching those with ADHD.

Zentall & Zentall (1983) identified the Optimal Stimulation Theory, which "defined activity as a regulatory mechanism for the maintenance of optimal sensory input;" activities linked with distractibility and the quest for stimulation can include "increased locomotor activity, looking around, and increased verbalization" (p. 453). This theory is critical for those having ADHD because the theory included the control or manipulation of arousal through activity. Abikoff, Courtney, Szeibel, & Koplewicz (1996) posit that distractibility of children with ADHD may be a way to compensate for underarousal as they search for "stimulation or novelty" through a variety of activities such as looking around or excessive movement (p. 238-9).

One additional model encompassing both arousal and dopamine release is the

Moderate Brain Arousal (MBA) model (Sikstrom & Soderlund, 2007). Stochastic resonance, a phenomenon in which a stimulus that would typically be undetected by an observer becomes detectable or noticeable when another stimulus—noise—is added (p. 1053), is crucial to this model. Similar to the Yerkes-Dodson Law of Arousal, moderate noise levels allow for the best detection of the previously undetected stimulus while strong or weak noise levels caused detection abilities to decline. "...The peak of the [stochastic resonance] curve depends on the dopamine level, so that participants with low dopamine levels (ADHD) require more noise for optimal cognitive performance compared to controls" (Söderlund, Sikström, & Smart, 2007, p. 844). Söderlund, Sikström, & Smart (2007) compared white noise to silence as a background for a "high memory performance task, and a verbal task (VT), as a low memory task" (p. 840). Participants with ADHD performed better on the cognitive tasks in the white noise condition while their typical peers performed better on the cognitive tasks in silence. Hence adding sound to the background adds novelty to a situation and may also increase cognitive functioning by raising dopamine levels in those diagnosed with ADHD. Because reactions to sound stimuli appear to differ between the typical population and those diagnosed with attention deficits, the next section reviews literature using background music with specialized populations.

Background Music Use with Specialized Populations

Several studies show the benefits found in background music use in students with Intellectual Disabilities (formally known as Mental Retardation or MR) and emotional and behavioral disabilities. Gregoire (1984) found a decrease in negative behaviors when a Paraguayan harp ballad in triple meter during a short rest period among participants with Intellectual Disabilities (ID) as compared with a silent rest period. Reardon and Bell (1970) found that "stimulative" music decreased self-stimulation behaviors in boys with severe ID as compared to a silent control, sedative music, and spoken recording over a period of 4 days; however, a consistent resurgence of self-stimulation behaviors on Day 3 of each sound condition may be due to novelty of the sound decreasing the self-stimulation behavior.

Savan found that listening to the works of Mozart had a positive effect on their behavior during 70-minute science lessons (1998) and showed an increase in coordination and positive behaviors while exhibiting a significant decrease in blood pressure, body temperature, and pulse (1999) among students with emotional and behavior disorders. In both studies, the reduction of the physiological parameters correlated with more positive behaviors in the classroom such as increased concentration, fewer attention-seeking behaviors, lower noise level, and task completion. Hallam and Price (1998) noted a significant increase in math problem accuracy among the four calming music sessions as compared with the four non-music sessions among ten students with emotional and behavioral difficulties. The researchers reported that "the pupils who benefited most were those whose difficulties were associated with constant stimulus seeking and over-activity, closely resembling the 'hyperactive' syndrome" (p. 90), a term used previously to describe those with ADHD.

Several researchers have conducted research into the benefits of background music among students with attention deficits during a variety of activities. Klien (1981) found that slower tempo music allowed 65% of hyperactive participants to have results on a motor task measuring speed and accuracy closer to that of the typical population; all participants were most inaccurate with the fast tempo music. Cripe (1986) used rock and roll music to assess activity level among eight boys, ages 6 through 8, with attention deficits. Each participant was tested individually in a playroom while listening to music or silence through a set of headphones. A significant decrease in activity level was noted while listening to music as compared to the no music condition.

Academic-based tasks have also be studied. Scott (1970) used music from the Beatles as a background while four students with hyperactivity solved math problems. Three of four participants showed a significant increase in math problems completed and problems answered correctly in the "normal" classroom condition with background music. Abikoff, Courtney, Szeibel, and Koplewicz (1996) used three sound conditions: background music based on individual preferences with 98% choosing rock or rap music, background speech (an audiotape of the business report recorded from the local television broadcast), and silence while boys with and without ADHD completed an arithmetic task. Participants were administered the math tests in 30-minute sessions with a different sound condition every 10 minutes; sound conditions were rotated to control for order effect. Students with ADHD receiving the music condition first had significantly more correct answers than all other participants, including the nondisabled children (p. 242). Wiebe (2007) found rock music used by an adolescent with ADHD assisted with concentration on homework; other benefits included increased comprehension, and a more positive mood when listening to his favorite music. The boy frequently used music to regulate his mood, which helped him persist in his schoolwork. Similarly, Abikoff et al. (1996) noted "level of appeal, rather than the mere presence of stimulation, appears to be the critical influential feature for children with ADHD" (p. 243).

The effect of background music has been studied in combination with electroencephalograph (EEG) neurofeedback training, which has shown success as an alternative therapy for those diagnosed with ADHD with a self-regulation emphasis to improve focus, concentration, impulse control, mood control, relaxation, and on-task behavior. Pratt, Abel, and Skidmore (1995) used EEG neurofeedback training at a rate of three sessions per week with nineteen subjects ages 6 to 17 classified as ADD or ADHD for a total of thirty-nine sessions. Participants used a computer game to change their brainwave patterns, with an experimental group listened to the music of Mozart during sessions. All subjects reported post-treatment improvement; significant improvement only existed in participants with ADD who listened to Mozart. Three of the participants listening to music may have been coordinating their movements, known as entraining, to the music's rhythm "during the theta bar reduction task" (p. 31). Pratt, Abel, and Skidmore (1995) called for similar studies with "focus exclusively on one variant of attention deficit disorder" and larger samples (p. 31).

Potential benefits have been found among those using background music during various activities, particularly those diagnosed with attention deficits. No studies have been completed focusing on a specific indicator of ADHD while using background music, nor have any studies researched college students having ADHD. The purpose of this study was to investigate the effect of music listening on impulsivity as judged by the Conners' Continuous Performance Task (CPT) II v. 5. The specific research question addressed was if a significant difference exists in the level of impulsivity among participants as judged by the Conners' Continuous Performance Task (CPT) between an initial condition, a music condition, and a silence condition.

Methodology

Study participants were college undergraduate students between the ages of 18 and 24 both with and without attention deficits; participants were recruited from a large university in the Southwestern United States. GPower 3.1.5 Freeware (1992-2012) was used to ascertain sample size *a priori* using design, intended analysis, effect size f = .25, and power = .80. Although 14 participants were recommended for each group based on calculations, it was determined to use 30 in each group due to the Central Limit Theorem (Gravetter & Wallnau, 2013, p. 209).

Recruitment advertisements were made using flyers posted in residence halls, departments known to assist students with learning difficulties, and any other buildings that allowed flyers promoting research recruitment on campus after obtaining written permission from appropriate parties. Craigslist, Facebook, and the University Master Calendar were also utilized. The advertisement notified interested students they would be paid $10 cash to participate in the study.

Potential participants emailed the researcher; a consent form was sent upon this initial contact to allow for informed consent. These individuals were asked to provide their age, gender, confirmation of attention deficits, and several options for appointment times after reading the consent form as confirmation of intent to participate. Students were not required to produce evidence of diagnosis or treatment for attention deficits; any participant told by a medical professional or an educator that testing was recommended qualified as having a history of attention deficits regardless of treatment history. No students under the age of 18 were recruited per IRB protocols; individuals who would soon turn 18 were invited to email the researcher again after their birthday in the event that participants were still needed. Recruitment was extended twice to attain full enrollment of the typical needs group (*n* = 30, females = 18, males = 12, mean age = 20.47). Only two volunteers were recruited in the attention deficit group by the end of the second extension. Modifications to simplify flyer wording, and protocol change involving a reminder email sent to potential participants approximately 48 hours after dispersing the initial recruitment information were both approved by IRB before implementation. After an unsuccessful summer recruitment, fall recruitment was successful in acquiring students with attention deficits between the beginning and midterm of the fall semester (*n* = 29, females = 16, males = 13, mean age = 20.41; *N* = 59). Attrition and technological difficulties resulted in decease in the total sample size, with the final sample including typical group *n = 26* and *n* = 25 group with attention deficits (*N* = 51). Information collected with the intake survey can be found in Table 1. Table 2 contains subtype information as reported by participants in the ADHD group; self-reported comorbidity for both groups is shown in Table 3. The typical group contained five participants did report diagnoses for anxiety or depression, but this group was not required to report other conditions. Table 4 contains mean T-scores and Standard Deviations for attention deficit behavior categories collected using the Conners' Adult ADHD Rating Scales (CAARS). These scores were collected to show the general differences in ADHD behavior indicators between the groups. All mean scores were consistently elevated in the ADHD group as compared to the typical needs group with a mean difference ranging between 8.31 and 19.17 except for Self-Concept scores (mean difference = 3.42).

Procedure

All testing protocols were established in cooperation Dr. Jaswinder Ghuman, a psychiatrist specializing in attention deficits, and peer reviewed by at least one of Dr. Ghuman's colleagues. The protocols were tested in a pilot tested and approved on February 18, 2013 by the Institutional Review Board at the University of Arizona, IRB Project 13-0112.

The group with typical needs was tested during the semester in which they were recruited to avoid attrition. After group recruitment was completed, individual appointments were made with each participant. Each participant was randomly assigned a sound condition order using stratified random sampling; stratification was used to allow equal numbers of students with and without attention deficits in each condition order and to avoid confounding factors of fatigue and order effect. Participant using attention deficit-related medication were asked to skip the medication on the appointment day.

Participants were given an intake survey to obtain general information such as gender, age, and type of attention deficit present if applicable. The survey was also designed to ensure compliance to protocols regarding medication for those with attention deficits. All participants were also reminded of their rights as participants and invited to ask any questions about the paperwork or the study. Participants were administered the Conners' Continuous Performance Test (CPT) a total of three times; participants were asked if questions regarding the study existed prior to each test administration.

The researcher remained in the room for the initial administration, which served as a control for the confounding effect of practice. The participant was allowed to ask questions and became familiar with the CPT before administration in the music and the silence conditions.

The remaining conditions of music and silence were administered in a separate space from the control administration. The researcher left the testing area after preparing the CPT and establishing the sound condition; however, participants were told to ask for help in the event of technical issues. Per the suggestion of Dr. Ghuman, each participant was given a short break and provided a small snack before the final CPT administration outside of the formal testing area to avoid the confounding factor of fatigue. The snack and break gave the participants the opportunity to revitalize slightly due to the large amount of energy required to focus on two 14-minute tests. The CAARS assessment was administered during the break and snack period. After the break, the final CPT was administered in the formal testing area using the other sound condition. Below is the protocol in minutes for each participant.

5 minutes	Introduction, verification of paperwork, brief survey
14 minutes	Participant took initial Conners' CPT on a laptop • Informal setting, such as researcher's office • Researcher present
5 minutes	Moved to formal setting; condition set up (silence or music condition) • For music condition, music was played from sound system separate from the computer. Researcher selected a piece of music that had a tempo around 120 and "looped" the music to ensure it played for the duration of the test. • Same laptop computer used • Room door was left ajar so researcher can observe; subject was instructed to notify researcher if technical issues arose
14 minutes	Participant repeated Conners' CPT in silence or music condition
15 minutes	Snack; Administration of the Conners' CAARS survey; received compensation of $10 cash and signed sheet stating compensation received
14 minutes	Participant repeated Conners' CPT in other sound condition (if silence first, than music or vice versa)
Total time: 67 minutes	

Instruments

In addition to the intake survey described above, the Conners' Adult ADHD Rating Scales (CAARS) (Conners, Erhardt, & Sparrow, 1998) were used to collect self-reported indicators of ADHD. CAARS (Conners et al., 1998) self- report instrument, long version was used as a screening tool to obtain a general profile of all participants; the long version was used in an attempt to collect the most accurate profile possible. The form contains 66 items and addresses symptomology as found in the *Diagnostic and Statistical Manual of Mental Disorders (DSM-IV)* and "factor-derived subscales" including "inattention/memory problems, hyperactivity/restlessness, impulsivity/emotional lability, and problems with self-concept" (Conners, Erhardt, and Sparrow, 2012, "Product Overview"). This allowed for group comparison based on typical behavioral categories of ADHD; each form was scored using the scoring grid and converted to T-scores based on norm group comparison. The scores were useful in that the potential presence of attention deficits were minimally corroborated if formal diagnosis had not been obtained. Response patterns and T-scores provided a profile of those participants not having a formal diagnosis, which were reviewed to determine whether to include the respective CPT scores in the final analysis. Because the CPT was administered using a computer, the CAARS was administrated on paper as opposed to an online administration version to provide participants a computer "screen break."

Conners' Continuous Performance Test (CPT) II (Version 5) for Windows, a signal detection program, was used to measure impulsivity levels of the participants. During the CPT, letter stimuli are displayed at one, two, and four second intervals, with a total administration time of 14 minutes. Upon seeing the letter stimuli, participants were to press the spacebar unless an X appeared on the screen. The participant was told to avoid pressing the spacebar when seeing an X. The CPT software compared the participant's results to the test's normative group, which were then converted to T-scores. The score reports were downloaded and transferred to an Excel spreadsheet at the end of the study.

Materials

The song *In a Mellow Tone* by Count Basie was played in the background during the music condition. This piece was chosen for testing as it was likely to be unfamiliar to the participants, the work had familiar instrumentation such as a drum set, and jazz music typically has not been associated with the Mozart Effect. The drum set also ensured the consistent presence of the beat in this musical selection. For administration purposes, the researcher used Audacity 1.3 Beta (software program) to manipulate the original tempo of the piece (approximately mm = 98) to be approximately mm = 120 (actual tempo mm = 124). The abbreviation mm refers to beats per second; a tempo of mm = 60 would have 1 beat per second whereas a tempo of mm = 120 would have 2 beats per second. The tempo was altered due to the researcher's previous experiences with students having attention deficits; students with attention deficits were observed using background music around this tempo prior to this study.

Hence the tempo of mm = 124 seemed a reasonable equivalent. The piece was then looped ensure the music lasted without interruption for 14 minutes, which was the length of the CPT administration. The loop was strategically created so that the loop made musical "sense" to avoid an "obvious" break for the listener. A sound system (iPod and a stand-alone speaker system or a CD player) separate from the laptop used to administer the CPT was used to play the music. Volume levels were set at the beginning of testing by the researcher; participants were then asked if the volume level aligned with personal comfort due to potential differences in hearing ability. All participants either confirmed the volume level as being comfortable or requested minor adjustments (i.e. "just a little softer"). The dial was returned to its original position if adjustments were made to the volume upon the completion of the sound condition.

A HP 2420 Netbook provided by the researcher was used to administer the CPT, and the same netbook was used for all test administrations. Due to a lower profile spacebar, it was found during the pilot study participants were not fully depressing the spacebar on the netbook. Hence, a Microsoft Wireless Comfort Desktop 5000 keyboard was used during this study instead of the netbook keyboard because the external keyboard was found to be more sensitive.

Design

Because it was unknown if the ADHD group would respond similarly to the sound stimuli as the typical group, the initial condition with no or minimal treatment was suggested in the protocol design process and supported in the literature (Chambless & Hollon, 1998; Saville & Buskist, 2005). The initial condition was included as part of the analysis and used as a point of comparison to determine the effects of sound conditions on impulsivity levels; the initial condition was also utilized to confirm homogeneity between the typical and ADHD groups. The initial condition allowed participants to experience the CPT and ask questions before administration in the two formal test administrations.

Commissions, Hit Reaction Time, and Perseverations were impulsivity measurements provided by the CPT software; a continuous scale was used to determine impulsivity outcomes.

Assessment criteria are defined below based on the definitions in the CPT Technical Manual for each impulsivity measure (Conners & MHS Staff, 2004):

- Commissions: Number of responses to non-target (X)
- Hit Reaction Time: Mean response time in milliseconds for all target responses in all six blocks, which are divisions in place to see consistency over time
- Perseverations: A response occurring less than 100ms following a stimulus

A resultant Two-Way Analysis of Variance (ANOVA) with Repeated Measures was computed using CPT Commission scores as the dependent variable and Group and Condition as the independent variables to determine whether a significant difference existed in impulsivity scores (CPT Commission scores) by Group (Typical versus ADHD groups) and by Conditions (Initial versus Music versus Silence Condition).

Results

Data were collected on a sample of $N = 51$, with an $n = 26$ for the typical group and an $n = 25$ for the ADHD group. Due to normality issues with few of the measures passing the Shapiro-Wilk or the Kolmogorov-Smirnov normality tests, T-scores calculated by the CPT were used for the analysis as opposed to raw scores as the dependent variable measure. The Winsorization procedure outlined by Howell (2013) was used to remove any detected outliers from the Commissions Pretest measures with 15% Winsorization of scores. No outliers were detected in either the music or the silence condition scores. All Commissions measures met the criteria for skewness, kurtosis, and normality Winsorization of the Commission pretest measures explained above. Results for these tests are displayed in Table 5. There were three potential CPT measures that could have been used as dependent variables: Commissions, Hit Reaction Time, and Perseverations. High correlations in these three CPT measures required further analysis to determine whether to include all 3 measures in the analysis due to multicollinearity concerns. Bivariate correlations between initial condition scores were used to quantify the relationships between the CPT measures; Commissions were significantly correlated with both Hit Reaction Time and

Perseveration measures (see Table 6). Regression was used to reveal high Squared Multiple Correlations (SMC)—calculated from Tolerance measures—with values ranging from .50 to .84 (see Table 7). Tabachnick and Fidell (2013) stated that covariates with SMC values higher than .5 may be eliminated from further analysis due to high correlation with the dependent variable(s) (p. 203). Therefore, the CPT measures Hit Reaction Time and Perseveration were determined to be assessing the same thing as the CPT measure Commissions. Thus, Hit Reaction Time and Perseveration were removed from the analysis.

In addition to normality, data were screened for accuracy and tested for order effect, linearity, and homoscedasticity. Order effect was analyzed using a One-Way ANOVA with repeated measure. No significant differences were found with F (1, 49) = .04, p = .84 between the randomized conditions of music (M = 60.28, SE = 2.09) and silence (M = 59.70, SE = 2.13). Bivariate linearity was analyzed using scatterplots with no detectable patterns found in the residuals. Homoscedasticity was confirmed using Box's M (p = .07) and Levene's Test with p = .25 for the initial condition, p = .89 for the music condition, and p = .47 for the silence condition. All dependent variables were found to be independent with sphericity validated using Mauchly's W (p = .43).

A significant difference was revealed upon analysis of a Two-Way ANOVA with repeated measure with F (1, 49) = 6.42, p = .02, d = .72, 95% CI [-12.73, -1.47] between the typical group (M = 56.52, SE = 1.96) and the ADHD group (M = 63.62, SE = 2.00) with a mean score difference of 7.1 and medium to large treatment effects. The typical group exhibited lower average T-scores in all conditions than the ADHD group in that the typical group "hit" fewer Xs (i.e. non-targets), which equates lower impulsivity and better performance as evidenced by the Commission mean scores.

The analysis above also revealed a significant difference between sound conditions with F (2, 98) = 8.22, p = .001, d = .82 with a large treatment effect among conditions. Further analysis of pairwise comparisons with Bonferroni correction showed initial condition mean scores (M = 57.46, SE = 1.31) were significantly lower than both the music condition mean scores

(M = 60.77, SE = 1.69) with a mean difference = -3.30, p = .03, 95% CI [-6.41, -.20] and the silence condition mean scores (M = 61.97, SE = 1.63) with a mean difference = -4.51, p = .001, 95% CI [-7.30, -1.72]. No significant mean score difference were found between the music condition and the silence condition (mean difference = -1.21, p = .80, 95% CI [-3.86, 1.45]). Participants in both groups performed best during the initial condition in that T-scores were significantly lower in the initial condition while T-scores in the music and the silence condition were statistically equivalent. Both the music and the silence conditions had higher Commission mean scores than the initial condition, which equated to a poorer performance among participants in the typical and the ADHD groups.

No significant interaction was found between the factors of group and condition with F (2, 98) = .66, p = .52, d = .23 with small treatment effect. The Commission score mean differences between groups were not significant across the initial, music, and silence condition, and the factors of group and condition appear to be independent.

Discussion

Data were collected to determine whether playing music in the background had a significant effect on impulsivity in those with ADHD as judged by performance on a Continuous Performance Test (CPT), which is a repetitious computer task used as part of the ADHD diagnosis process. Mixed results have been reported by researchers regarding the use of background music during a variety of tasks; few of these studies attempted to measure the effect of background music on specific populations, with a limited number of these studies specifically focusing on students with attention deficits. This study was the primary study to ascertain the effect of background music on college students with ADHD and to assess the effect of background music on one specific indicator of ADHD.

The data were analyzed to determine whether significant differences existed between a typical group (participants with no history of ADHD) and an ADHD group (participants with a history of attention deficits in some way), among conditions (an initial condition with the researcher was present during the CPT and questions allowed by the participant, a background music condition in which jazz music was played during the CPT administration, and a silence condition in which the participant took the CPT in silence), and between groups by condition. Significant differences were found by group in Commission mean scores, with the typical group achieving lower Commission mean scores. Lower Commission scores during the 14-minutes test indicated better performance as fewer nontargets (Xs) were hit, which equated to lower impulsivity levels. Because the CPT can be used for diagnostic purposes, higher mean Commission scores were expected of those with ADHD as higher scores help differentiate those with ADHD from the typical population (Conners & MHS Staff, 2004).

Significant differences were also found by condition. All participants had significantly lower Commission mean scores in the initial condition as compared to the music and the silence condition, which equated to lower impulsivity and better performance. No significant Commission mean scores difference were found between the music and silence conditions, but mean scores from both conditions were higher as compared to the initial condition. It is possible that level of arousal could account for these differences.

Each participant arrived in the office space, finished the intake survey, and began the CPT immediately thereafter with the researcher present. Mehrabian (1977) stated that higher arousal is exhibited during sudden or surprising situations with lower arousal during more calm or monotonous conditions. The presence of the researcher in combination with an unfamiliar space and testing procedure may have caused heightened arousal in participants' to a moderate level. Heightened arousal positively affected participant performance in that all participants were more focused and less impulsive as judged by CPT Commission scores.

For the ADHD group specifically, it could be hypothesized that initial condition Commission mean scores were lower due to higher levels of arousal during the initial test administration as compared with scores in the music and silence condition. This hypothesis would be consistent with the Optimal Stimulation Theory (Zentall & Zentall, 1983). Abikoff, Courtney, Szeibel, &

Koplewicz (1996) posit that "stimulation or novelty" may be a way to compensate for underarousal as part of the Optimal Stimulation Theory, which leads to distractibility of children with ADHD (p. 238). The initial condition in an unfamiliar setting with an unfamiliar test would be a novel situation that would quickly lose its uniqueness after repeated administrations. The playing of background music during the administration of the CPT did not enhance arousal to the extent that Commission mean scores were lower in the music condition than in the initial condition, and the Commission mean scores in the music condition were statistically equivalent to Commission mean scores from the silence condition. Reardon and Bell (1970) experienced a similar confound when studying the self-stimulation behaviors of boys with severe intellectual disabilities; it was uncertain whether the decrease in self-stimulation behaviors was due to the novelty of having sound in the room or specifically the music provided.

Music preference has been found to affect the arousal level, with researchers finding preferred music positively influence mood and states of arousal (Lesiuk, 2010), stress reduction (Jiang et al., 2013), and results of an arithmetic task (Abikoff et al., 1996). Although some researchers have found some benefits using the music of Mozart to those with special needs, including those with ADHD (Pratt, Abel, and Skidmore, 1995; Savan, 1998; Savan, 1999), others have found benefits to those with ADHD using rock music (Abikoff et al., 1996; Cripe, 1986; Scott, 1970; Wiebe, 2007). Still others have found music to either be a distraction or have obtained mixed results among participants when played in the background (Anderson & Fuller, 2010; Fogelson, 1973; Freeburne & Fleischer, 1952; Henderson, Crews, & Barlow, 1945). Because arousal levels are individualized, the music selection played may not have been able to raise the arousal levels of all participants consistently during the music condition. Hence genre and preference could be factors to the equivalence of Commission mean scores in the music and silence condition, which could explain the music condition mean scores being equivalent to those in the silence condition.

Because the factors of group and condition appear to be independent, the Commission mean score differences were not significant by group across the three sound conditions. No interaction was found.

Study Limitations

Study limitations included the use of a participant self-report regarding history of attention deficits as well as the use of a convenience sample. Although the office environment used for this study was relatively removed from high traffic areas, the environment became a limitation to this study. Using an iPod to provide music during the ADHD group testing also became problematic. These limitations outlined in more detail below.

Participants for the ADHD group were not required to have clinical diagnosis; rather, participants were identified through a self-report form with no corroboration from other observers. This resulted in confusion for two participants who were compensated for their time but were removed from the data set. The confusion stemmed from self-perceived attention deficits (i.e. "I feel I have difficulty focusing") as opposed to a history (i.e. "I was encouraged to obtain further evaluation for observed issues"). Although the remaining ADHD sample obtained scores on both the CAARS rating scales and the CPT consistent with those formally diagnosed with ADHD, further corroboration with another observer—similar to requirements for a clinical diagnosis—or requiring a clinical diagnosis to participate would have strengthened the study.

Another limitation was use of a convenience sampling procedure rather than a random sampling procedure. The probability of selecting a representative sample from the population is highly unlikely when utilizing volunteers as opposed to random selection from the population means. A random sample should be a consideration for future studies as different results are possible.

Although a soundproof room would have been ideal, this study was conducted in an office environment. Traffic in the office area was minimal, but the environment was not completely silent (i.e. hall noise, sounds from a neighboring office, etc.). Any sounds heard in the testing area could have caused distractions and affected participant performance. Distracted participants may have hit more nontargets, which results in high Commission mean scores. This is especially true in the silence condition as the condition may not have been truly silent. A soundproof room for this study, although ideal, was not an option.

Two CD players stopped working in the middle of testing sessions during the typical group testing period. To prevent further difficulties, a speaker with an iPod was used for the attention deficit group. Although the iPod screen darkens shortly after the play button is depressed, it was possible to see the amount of time elapsed. Several participants mentioned attempts to see how much time remained in the test as the iPod was visible from the testing area. This information may have affected the participants in either positive ways (i.e. this monotonous task is almost over) or negative ways (i.e. missing targets on the screen while attempting to see the time on the iPod).

Suggestions for Future Research and Implications

Consideration should be given to a mixed methods study involving both quantitative measurements and qualitative statement analysis. Although it was not formally included in the design, participants in the typical group anecdotally stated focusing was difficult in the music condition after test completion. Participants in the attention deficit group had a different reaction, typically stating that focusing was easier in the music condition. By analyzing comments of the participants, possible correlations may emerge between positive or negative comments and actual CPT performance. It was also noted by the researcher that participants were depressing the spacebar with the music's beat as the spacebar click was audible from where the researcher was sitting outside of the room. The act of depressing a spacebar to the beat of music is an act of entrainment—a coordinated rhythmic movement to a pulse. Further investigation could also include potential entrainment effects.

A randomly selected sample would also be recommended as would recruiting a sample based on diagnostic subtype. Although this study's sample was large enough for the proposed analysis, a more detailed sample selection may be necessary to truly show potential benefits for those having ADHD. Ideally, recruiting specifically by diagnostic subtype (i.e. hyperactive/impulsive, inattentive, and combined type) would be

necessary as the characteristics of each subgroup are very different. Hence, the effects of music on different diagnostic subtypes may also be different. Although subtype information was collected as part of the intake process, the information was not consistently available and was not examined in this study. In addition, the sample for the current study was not randomly selected, which limits generalizability because results are likely to be different using a random sample.
Although music preference was a large factor in at least two studies, this study did not examine music preferences of the participants. Lesiuk (2010) found participants remaining focused, calm, and relaxed during high-cognitive demand tasks when listening to preferred music. Abikoff et al. (1996) posited "... the level of appeal, rather than the mere presence of stimulation, appears to be the critical influential feature for children with ADHD" (p. 243). Future research could consider the effect of music preference on task completion, especially in repeated or "unpleasant" tasks. Another possible factor to consider is arousal; Robb (2003) stated that music can have an impact on arousal. Future research could contemplate the roles of music preference and arousal during a variety of tasks. Additional research is required to see if music preference would improve or decline scores of participants with and without attention deficits.

In addition, further investigation is needed to determine the effect of background music use during academic tasks. Researchers could explore if background music has an effect on students completing a variety of educational tasks (i.e. math problems, reading, homework completion, etc.), especially if students are accustomed to having background sound in some form while completing these activities in their typical environment.

It is unknown what style music might have the most impact with students diagnosed with ADHD due to the scant literature base. Music education has a role in this research as these professionals are the musical guides of their respective schools, especially in the general music setting. Because technology allows students to listen to isolated music genres, music teachers help students discover and explore genres of which they may not be aware. Music educators should discuss potential effects of music or what musical elements could have an effect with their students. This may help students make conscious decisions regarding their music use.

This was the first study to investigate the effect of music on one specific indicator of attention deficits, as well as the first study to investigate the effect of background music on college students with attention deficits. A significant main effect difference was found by group with the typical group performing better than the ADHD group based on Commission mean scores. Only the initial condition was found to be significantly lower than both the music condition and the silence condition for both groups; the music and silence condition were statistically equivalent. The factors of group and condition were independent with no interaction found. Much work is needed to see if these findings can be replicated and continuing analysis on this data set for other indicators of ADHD.

References

Abikoff, H., Courtney, M. E., Szeibel, P. J., & Koplewicz, H. S. (1996). The effects of auditory stimulation on the arithmetic performance of children with ADHD and nondisabled children. *Journal of Learning Disabilities, 29*(3), 238-246.

American Psychiatric Association. (2013). *Diagnostic and statistical manual of mental disorders: DSM-5*. Washington, D.C.: American Psychiatric Association.

Anderson, S. A., & Fuller, G. B. (2010). Effect of music on reading comprehension of junior high school students. *School Psychology Quarterly, 25*(3), 178-187.
Audacity (Version 1.3 Beta) [Computer Software]. Boston, MA: Free Software Foundation, Inc.

Chambless, D. L., & Hollon, S. D. (1998). Defining empirically supported therapies. *Journal of Consulting and Clinical Psychology, 66*(1), 7-18.

Conners' Continuous Performance Test (CPT) II (Version 5) [Computer software]. Toronto, Ontario, Canada: Multi-Health Systems, Inc.

Conners, C. K., Erhardt, D., & Sparrow, E. (1998). *Conners' Adult ADHD Rating Scales (CAARS) Self-Report: Long version*. North Tonawanda, NY: Multi-Health Systems, Inc.

Conners, C. K., Erhardt, D., & Sparrow, E. (2012). CAARS: Conners' Adult ADHD Rating Scales: Product overview. Retrieved October 3, 2012, from https://ecom.mhs.com/(S(43xyb0ipbecg2j2tuy14pi55))/product.aspx?gr=cli&prod=caars&id=overview

Conners, C. K. & MHS Staff. (2004). *Conners' Continuous Performance Test (CPT) II for Windows: Technical guide and software manual*. North Tonawanda, NY: Multi-Health Systems, Inc.

Cripe, F. F. (1986). Rock music as therapy for children with Attention Deficit Disorder: An exploratory study. *Journal of Music Therapy, 23*(1), 30-37.

Fogelson, S. (1973). Music as a distractor on reading-test performance of eighth grade students. *Perceptual and Motor Skills, 36*(3c), 1265-1266.

Freeburne, C. M., & Fleischer, M. S. (1952). The effect of music distraction upon reading rate and comprehension. *Journal of Educational Psychology Journal of Educational Psychology, 43*(2), 101-109.
GPower (3.1.5 Beta) [Computer Software]. Germany: Freeware.

Gravetter, F. J. & Wallnau, L. B. (2013) *Statistics for the Behavioral Sciences*. Belmont, CA: Wadsworth Cengage Learning.
Gregoire, M. (1984). Music as a prior condition to task performance. *Journal of Music Therapy, 21*(3), 133-145.

Hall, J. C. (1952). The effect of background music on the reading comprehension of 278 eighth and ninth grade students. *Journal of Educational Research, 45*(6), 451-458.

Hallam, S., & Price, J. (1998). Can the use of background music improve the behaviour and academic performance of children with emotional and behavioural difficulties? *British Journal of Special Education, 25*(2), 88-91.

Hallam, S., Price, J., & Katsarou, G. (2002). The effects of background music on primary school pupils' task performance. *Educational Studies, 28*(2), 111-122.

Henderson, M. T., Crews, A., & Barlow, J. (1945). A study of the effect of music distraction on reading efficiency. *Journal of Applied Psychology Journal of Applied Psychology, 29*(4), 313-317.

Howell, D. C. (2013). *Statistical Methods for Psychology* (8th ed.). Belmont, CA: Wadsworth Cengage Learning.
Jiang, J., Zhou, L., Rickson, D., & Jiang, C. (2013). The effects of sedative and stimulative music on stress reduction depend on music preference. *The Arts in Psychotherapy, 40*(2), 201-205.

Kämpfe, J., Sedlmeier, P., & Renkewitz, F. (2011). The impact of background music on adult listeners: A meta-analysis. *Psychology of Music, 39*(4), 424-448. doi: 10.1177/0305735610376261

Klein, P. S. (1981). Responses of hyperactive and normal children to variations in tempo of background of music. *The Israel Journal of Psychiatry and Related Sciences, 18*(2), 157-166.

Kotsopoulou, A., & Hallam, S. (2010). The perceived impact of playing music while studying: age and cultural differences. *Educational Studies, 36*(4), 431-440.

Lesiuk, T. (2010). The effect of preferred music on mood and performance in a high-cognitive demand occupation. *Journal of Music Therapy, 47*(2), 137-154.

Mehrabian, A. (1977). Individual differences in stimulus screening and arousability. *Journal of Personality, 45*(2), 237-250.

Ornitz, E. M., Gabikian, P., Russell, A. T., Guthrie, D., Hirano, C., & Gehricke, J. G. (1997). Affective valence and arousal in ADHD and normal boys during a startle habituation experiment. *Journal of the*

American Academy of Child and Adolescent Psychiatry, 36(12), 1698-1705.

Patton, J. E., Stinard, T. A., & Routh, D. K. (1983). Where do children study? *The Journal of Educational Research, 76*(5), 280-286.

Pratt, R. R., Abel, H.-H., & Skidmore, J. (1995). The effects of neurofeedback training with background music on EGG patterns of ADD and ADHD children. *International Journal of Arts Medicine, 4*(1), 24-31.

Reardon, D., & Bell, G. (1970). The effects of sedative and stimulative music on activity levels of severely retarded boys. *American Journal of Mental Deficiency, 75*(2), 156-159.

Robb, S. L. (2003). Music interventions and group participation skills of preschoolers with visual impairments: Raising questions about music, arousal, and attention. *Journal of Music Therapy, 40*(4), 266-282.

Savan, A. (1998). A study of the effect of background music on the behaviour and physiological responses of children with special educational needs. *Psychology of Education Review, 22*(1), 32-35.

Savan, A. (1999). The effect of background music on learning. *Psychology of Music, 27*(2), 138-146.

Saville, B. K., & Buskist, W. (2005). Traditional idiographic approaches: Small-N research designs. In S. F. Davis (Ed.*), Handbook of Research Methods in Experimental Psychology*. Blackwell Reference Online: Blackwell Publishing. Retrieved from http://www.blackwellreference.com.ezproxy2.library.arizona.edu/subscriber/tocnode.html?id=g9781405132800_chunk_g97814051328005.

Schellenberg, E. G., Nakata, T., Hunter, P. G., & Tamoto, S. (2007). Exposure to music and cognitive performance: Tests of children and adults. *Psychology of Music, 35*(1), 5-19.

Scott, T. J. (1970). The use of music to reduce hyperactivity in children. *The American Journal of Orthopsychiatry, 40*(4), 677-680.

Sikstrom, S., & Soderlund, G. (2007). Stimulus-dependent dopamine release in Attention-Deficit/Hyperactivity Disorder. *Psychological Review, 114*(4), 1047-1075.

Söderlund, G., Sikström, S., & Smart, A. (2007). Listen to the noise: Noise is beneficial for cognitive performance in ADHD. *Journal of Child Psychology and Psychiatry, 48*(8), 840-847.

Stanton, H. E. (1973). The effect of music on test anxiety. *Australian Psychologist, 8*(3), 220-228.

Tabachnick, B. G., & Fidell, L. S. (2013). *Using Multivariate Statistics*. Harlow: Pearson.

Wiebe, J. (2007). *ADHD, the classroom and music: A case study.* Master of Education Unpublished, University of Saskatchewan, Saskatoon.

Yerkes, R. M., & Dodson, J. D. (1908). The relation of strength of stimulus to rapidity of habit-formation. *Journal of Comparative Neurology and Psychology, 18*, 459-482.

Zentall, S. S., & Zentall, T. R. (1983). Optimal stimulation: A model of disordered activity and performance in normal and deviant children. *Psychological Bulletin, 94*(3), 446-471.

Table 1

Intake Information by Group

Group *N* = 51	Gender	Mean Age	Year in College	Formally diagnosed with ADHD	Currently taking medication for ADHD
Typical ***n* = 26**	Male = 10 Female = 16	20.58	Freshman = 6 Sophomore = 6 Junior = 6 Senior = 6 Fifth year = 2	Yes = 0 No = 26	Yes = 0 N/A = 26
ADHD ***n* = 25**	Male = 12 Female = 13	20.56	Freshman = 6 Sophomore = 3 Junior = 5 Senior = 6 Fifth year = 5	Yes = 19 No = 6	Yes = 11 No = 14

Table 2

ADHD Group and Self-Reported Subtype

Subtype	Frequency
Question left blank	6
Inattention	7
Hyperactivity	4
Impulsivity	1
Unsure	3
Inattention and Hyperactivity	1
Inattention and Impulsivity	2
All 3 Subtypes	1

Table 3

Self-Reported Comorbid Conditions

	Frequency by Group	
Comorbid Condition	**Typical**	**ADHD**
Blank/None of These	21	14
Anxiety	2	3
Depression	3	1
Learning Disability	0	3
Anxiety and Depression	0	3
Anxiety, Depression, and Learning Disability	0	1

Table 4

Mean T-scores and Standard Deviations of CAARS Categories by Group

	Inattention/Memory Problems		Hyperactivity/ Restlessness		Impulsivity/ Emotional Lability		Problems with Self-Concept	
Group	Mean	SD	Mean	SD	Mean	SD	Mean	SD
Typical	50.19	6.63	50.73	8.10	45.00	7.98	52.50	8.31
ADHD	63.56	11.62	59.04	11.42	54.88	11.02	55.92	13.66

	DSM-IV Inattentive Symptoms		DSM-IV Hyperactive-Impulsive Symptoms		DSM-IV ADHD Symptoms Total		ADHD Index	
Group	Mean	SD	Mean	SD	Mean	SD	Mean	SD
Typical	54.38	10.15	50.27	9.13	53.50	10.09	49.62	7.34
ADHD	73.56	13.41	62.68	10.61	71.80	12.01	61.40	10.96

Table 5

Normality Test Results for Commissions Measures

						Shapiro-Wilk		
Measure	Group	Mean	*SD*	Skewness	Kurtosis	Statistic	*df*	Sig.
Commissions Initial	Typical	54.21	8.92	.73	-.59	.951	26	.248
	ADHD	60.71	9.83	.32	-1.57	.929	25	.081
Commissions Music	Typical	57.68	12.08	.08	-.43	.983	26	.935
	ADHD	63.86	12.01	-.97	-1.05	.930	25	.087
Commissions Silence	Typical	57.67	12.65	.60	.12	.976	26	.784
	ADHD	66.28	10.44	-.31	-.81	.971	25	.662

Table 6

Initial Bivariate Correlations for CPT Measures

Measure	Commissions	Hit Reaction Time	Perseverations
Commissions	1.00	-.51	.51
Hit Reaction Time	-.51	1.00	.02
Perseverations	.51	.02	1.00

Note: $N = 51$ for this analysis

Table 7

Squared Multiple Correlation (SMC) Values as Compared to Commission Scores from the Initial Condition

Measures	SMC
Hit Reaction Time: Initial	.79
Hit Reaction Time: Music	.84
Hit Reaction Time: Silence	.84
Perseverations: Initial	.53
Perseverations: Music	.50
Perseverations: Silence	.59

A Web of Resonance: An Examination of Chinese Bell Therapy

Steve Jackowicz, Ph.D.

University of Bridgeport

Abstract

The Chinese *Bian Zhong* Bells are calibrated to the sexagenary cycle of traditional Chinese music theory which posits a relationship between the organs and meridians recognized in Chinese medicine, and the application of notes and sequences of notes in proximity to the body. Such notes and sequences were traditionally prescribed to change the state of health, both physical and emotional, following a specific schemata of interaction. In the Chinese medical text the *Yellow Thearch's Inner Canon (Huangdi Neijing)* there is a schema of interaction of notes in singular and combination that are used to affect the meridian system. Based on this traditional approach, a protocol of bell sequencing was developed and applied in an N20 research study. Physiological effects were observed in response to the use of the bell therapy on the test subjects indicating a potential for clinical application.

Introduction

Chinese Bell (Bianzhong) were historically used as not only a musical standard for Confucian musical scales, but also for therapeutic intervention. This study is based on previous research into the structure of the Chinese music therapy model (see Jackowicz, ISQRMM 2014 Proceedings Journal). The *Yellow Thearch's Inner Canon Spiritual Axis* Chapter 65 details a specific application of the bell system to address disorders of the meridian system used in Traditional Chinese Medicine. This study developed a protocol based on the method recorded in the *Spiritual Axis* Chapter 65 utilizing a redesigned bell tree. Twenty subjects were treated with the bell protocol and biometric results were recorded and tabulated showing change in response to exposure to the bell sound sequences.

Background

Historically the Chinese developed ornate theories of sound and its impact on the physiology. In the *Treatise of Music* (Yue Shu 樂書) in the *Historical Records* (Shi ji 史記) there is an ornate correspondence of sound to organic function:

> Thus Gong moves the spleen and harmonizes it to correct integrity,
> Shang moves the lungs and harmonizes it to correct righteousness,
> Jue moves the liver and harmonizes it to correct benevolence,
> Zhi moves the heart and harmonizes it to correct propriety,
> Yu moves the kidneys and harmonizes them to correct wisdom.

The Chinese notes are indicated by their names. Gong, shang, jue, zhi, and yu are equivalent to 1, 2, 3, 5, and 6 in numbered musical notation or do, re, mi, so, and la in western solfeggio. However the concept was that these sounds had a transcendent nodal quality and were emanations of the primary forces that made up the phenomenal world. As such they were correlated to the Five Phases (wood, fire, earth, metal, water) which enjoyed a pentiary relationship of mutual creation and restriction. The Five Phase model provided much of the philosophical basis for Chinese Medical theory interrelating physical tissue, moral and mental characteristics, and geomantic principles. In relation to the notes or nodal sounds the Five Phases relate as follows:

Five nodal sounds arranged with five phases

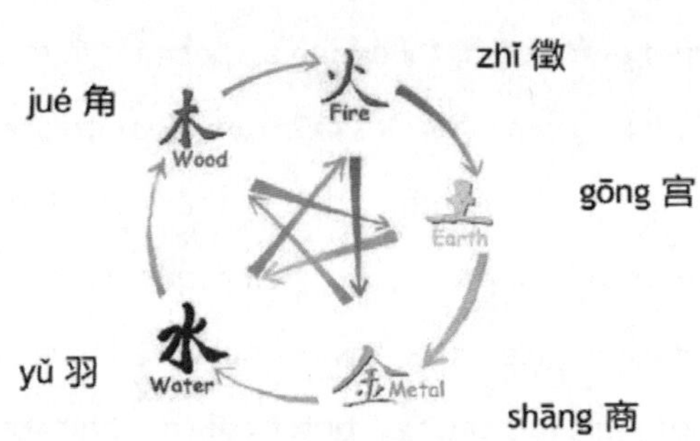

Chinese Medical thought further expanded the Five Phase Theory to organize herbal medicines, acupuncture, breathing practices, and even musical therapeutics.

The *Yellow Thearch's Inner Canon*, considered the preeminent text of classical medical theory has multiple entries detailing therapeutic application of music. Chapter 65 of the *Spiritual Axis* Section of the *Yellow Thearch's Inner Canon* details a specific approach to nodal sounds chimed in physical relation to the meridians of acupuncture on the body to create a therapeutic effect. Chapter 65 has:

> 右徵與少徵，調右手太陽上。左商與左徵，調左手陽明上。少徵與大宮，調左手陽明上。右角與大角，調右手少陽下。大徵與少徵，調左手太陽上。衆羽與少羽，調右足太陽下。少商與右商調右手太陽下。桎羽與衆羽，調右足太陽下。少宮與大宮，調右足陽明下。判角與少角，調右足少陽下。釱商與上商，調右足陽明下。釱商與上角，調左足太陽下。

> Right Zhi and soft Zhi regulate the right arm Taiyang Channel upper part. Left Shang and left Zhi regulates the left arm Yangming upper part. Soft Zhi and Loud Gong regulate the left arm Yangming upper part. Right Jue and Loud Jue regulate the right arm Shaoyang lower part. Loud Zhi and soft Zhi regulate the left arm Taiyang upper part. Resounding Yu and soft Yu regulate the right foot Taiyang lower part. Soft Shang and right Shang regulate right arm Taiyang lower part. Muted Yu and resounding Yu regulate the right Taiyang lower part. Soft Gong and loud Gong regulate right leg Yangming lower part. Clear Jue and soft Jue regulate the right leg Shaoyang lower part. Muted (Fettered) Shang and upper Shang regulate the right leg Yangming lower part. Muted (Fettered) Shang and upper Jue regulate the left foot Taiyang lower part.

> 上徵與右徵同穀麥、畜羊、果杏，手少陰藏心，色赤味苦，時夏。上羽與大羽，同穀大豆，畜彘，果栗，足少陰藏腎，色黑味鹹，時冬。上宮與大宮同穀稷，畜牛，果棗，足太陰藏脾，色黃味甘，時季夏。上商與右商同穀黍，畜雞，果桃，手太陰藏肺，色白味辛，時秋。上角與大角，同穀麻、畜犬、果李，足厥陰藏肝，色青味酸，時春。

> Upper Zhi and right Zhi both are of the grain wheat, the animal of the ram, the fruit apricot, arm Shaoyin Heart Orb, the color red, the flavor bitter and the season summer. Upper Yu and loud Yu both are of the grain of soy, the animal of the pig, the fruit chestnut, the leg Shaoyin Kidney Orb, the color black, the flavor salty, and the season of winter. Upper Gong and loud Gong are both of the grain of millet, of the animal cow, of the fruit dates, foot Taiyin Spleen Orb, the color yellow, the flavor sweet, and the season of late summer. Upper Shang and right Shang both are of the grain broomcorn millet, the animal chicken, the fruit peach, the arm Taiyin Lung Orb, the color white, the flavor

pungent, and the season Autumn. Upper Jue and loud Jue both are of the grain hemp, the animal dog, the fruit peach, leg Jueyin Liver Orb, the color green, the flavor sour, and the season spring.

大宮與上角，同右足陽明上，左角與大角，同左足陽明上，少羽與大羽同右足太陽下，左商與右商，同左手陽明上，加宮與大宮同左足少陽上，質判與大宮，同左手太陽下，判角與大角同左足少陽下，大羽與大角，同右足太陽上，大角與大宮同右足少陽上，右徵、少徵、質徵、上徵、判徵、右角、鈦角、上角、大角、判角。右商、少商、鈦商、上商、左商。少宮、上宮、大宮、加宮、左角宮。眾羽、桎羽、上羽、大羽、少羽。

Loud Gong and Upper Gong both affect right leg Yangming upper part. Left Jue and loud Jue both affect left leg Yangming upper part. Soft Yu and loud Yu both affect right leg Taiyang lower part. Left Shang and right Shang both affect left arm Yangming upper part. Additional Gong and loud Gong both affect left leg Shaoyang upper part. Plain clear and loud Gong all affect left arm Taiyang lower part. Clear Jue and loud Jue both affect left leg Shaoyang lower part. Loud Yu and loud Jue both affect the right leg Taiyang upper part. Loud Jue and loud Gong both affect the right leg Shaoyang upper part. Right Zhi, soft Zhi, plain Zhi, upper Zhi, clear Zhi, right Jue, Muted (Fettered) Jue, upper Jue, loud Jue, clear Jue. Right Shang, soft Shang, Muted (Fetterd) Shang, upper Shang, left Shang. Soft Gong, upper Gong, loud Gong, Additional Gong, left Jue Gong. Resounding Yu, Upper Yu, Loud Yu, Soft Yu.

There is a lack of material in the historical records to indicate how widely this method was practiced, or to indicate its efficacy. However it provides detailed evidence that music was used as a physical therapeutic interventive method as early as the third century BCE when the *Yellow Thearch's Inner Canon* was codified. Based on this approach, I developed a protocol to examine the impact of the Chinese Bells on the metabolism.

Method

The details of the Chinese Bell therapy from the *Spiritual Axis* included the following terms to describe the variations of the notes on the pentatonic scale:

Upper 上 shang

Loud 大 da

Soft 少 shao

Left 左 zuo

Right 右 you

Muted 桎 zhi – hit with the wooden part

Muted (Fettered) 鈦 di – stopped once hit

Clear 判 pan

Plain 質 zhi

Additional 加 jia – trembles when hit

Resounding 眾 Zhong

These notes are described as being sounded using the above variations in proximity to specific acupuncture meridians. Tabulating them yielded a chart as follows:

Note	Channel Affected	Side	Part
Right Zhi	Arm Taiyang	Right	Upper
Soft Zhi			
Left Shang	Arm Yangming	Left	Upper
Left Zhi			
Soft Zhi	Arm Yangming	Left	Upper
Loud Gong			
Right Jue	Arm Shaoyang	Right	Lower
Loud Jue			
Loud Zhi	Arm Taiyang	Left	Upper
Soft Zhi			
Resounding Yu	Leg Taiyang	Right	Lower
Soft Yu			
Soft Shang	Arm Taiyang	Right	Lower
Right Shang			
Muted Yu	Taiyang	Right	Lower
Resounding Yu			
Soft Gong	Leg Yangming	Right	Lower
Loud Gong			
Clear Jue	Leg Shaoyang	Right	Lower
Soft Jue			
Muted (Fettered) Shang	Leg Yangming	Right	Lower
Upper Shang			
Muted (Fettred) Shang	Leg Taiyang	Left	Lower
Upper Jue			

Note	Channel Affected	Side	Part
Loud Gong	Leg Yangming	Right	Upper
Upper Gong			
Left Jue	Leg Yangming	Left	Upper
Loud Jue			
Soft Yu	Leg Taiyang	Right	Lower
Loud Yu			
Left Shang	Arm Yangming	Left	Upper
Right Shang			
Additional Gong	Leg Shaoyang	Left	Upper
Loud Gong			
Plain Gong	Arm Taiyang	Left	Lower
Clear Gong			
Loud Gong			
Clear Jue	Leg Shaoyang	Left	Lower
Loud Jue			
Loud Yu	Leg Taiyang	Right	Upper
Loud Jue			
Loud Jue	Leg Shaoyang	Right	Upper
Loud Gong			

In order to have a portable bell mechanism that could be moved into proximal relationship to the meridian indicated in the chart, a bell tree was reassembled to only include bells corresponding to the Chinese pentatonic scale. These nodal sounds were then chimed in accord with the derived protocol.

Since all of the notes indicted in the source text are associated with the Five Phases, the protocol was organized around chiming the bell twelve times in the generative cycle of the Five Phases, then in the controlling cycle of the Five Phases, followed by an area specific use of the notes in relation to the acupuncture channels, as listed in the charts. This approach would theoretically impact the core organs of the body from the Chinese Medical perspective to enhance their functions, and then address the attendance surface concerns manifested in their channels.

Study

The protocol derived from the *Spiritual Axis* Chapter 65 was applied to a sample of 20 subjects, ranging from 18 to 37 years old, being 14 males and six females. Subjects were assessed for their transdermal skin resistance, thermography, and goniometer measures of all major joints (neck, shoulder, elbow, hip, and knee). Subjects were surveyed regarding sense of energy / fatigue, peacefulness / anxiety, pain / comfort. Subjects underwent a single treatment with a set Bell protocol and were reassessed

Results

The transdermal resistance dropped in all subjects. Mean drop equaled -10 millivolts. Thermal scans yielded a drop in surface temperature in all subjects. Mean drop equaled -0.8 degrees Fahrenheit. Goniometer readings improved in torso flexion for all subjects, in those with normal range of motion in peripheral joints there was no change, in those with abnormal range of motion there was an improvement in all subjects. Mean torso flexion improvement equaled 29 degrees increase.

Below is an example of the before and after thermographic scan of one subject:

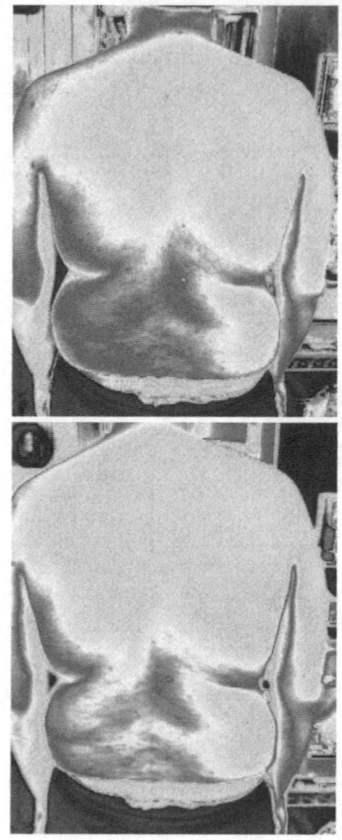

These scans show the drop in surface temperature on the back of the subject before and after Chinese Bell Therapy test.

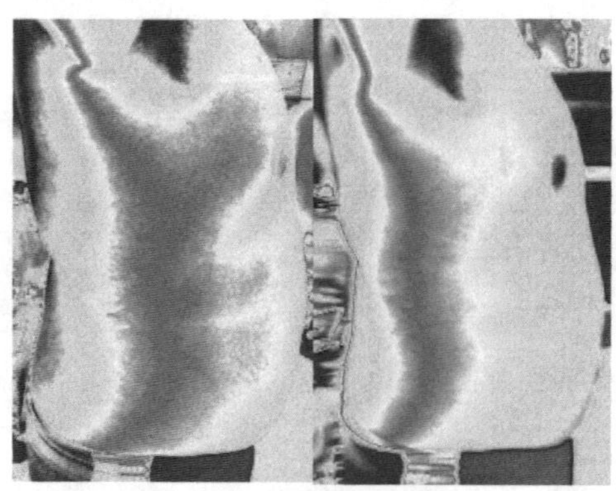

These scans show the change in surface temperature on the flank of the subject before and after the test.

Conclusion

This study is admittedly a pilot. However the changes in biometric parameters observed in the test implies that physiological changes are occurring in response to the bell therapy. There are limitations to the study in terms of the sample of subjects being of a large range in age, not being separated by gender, and not separated by any underlying pathologies. Further there is no way to blind the study to separate out any placebo effect. Despite these limitations the thermal change, and electrical resistance changes indicate that further examination of this protocol warrants attention for the potential to develop an understanding of its clinical application.

The Effect of Music Practice on Anxiety in Undergraduate Music Students

David O. Akombo,

Jackson State University, USA

Abstract

College students' lives are demanding thus exposing them to anxiety. This study was interested in finding whether anxiety levels would be affected by music practice. 25 undergraduate students above 18 years of age participated in the study. All the participants had a self-reported stress levels before and after the music practice. Their anxiety was measured by the State Trait Anxiety Inventory. It was hypothesized that the participants would have shown no significant differences after the half-hour practice. No significant relationship was found; thus, my hypothesis was supported.

INTRODUCTION

It is certainly not unusual for a college student who has worked long and hard in the week including moments of overnight class discussions and term papers to exhibit some form of anxiety. For the most part of the last three decades, scholars have had great interests in the use of music to control anxiety. Alleviation of stress anxiety was extensively addressed by Hammer (1996), who studied the relationship between music therapy and participants' perceived stress level. Participants participated in music therapy sessions that included relaxation techniques and guided imagery. The State and Trait Anxiety Inventory (STAI) was administered before and after the music therapy sessions. In the experimental group, there was a slight reduction in STAI levels and a perceived decline in stress levels after the music therapy sessions. These results suggest that music could be effectively used to reduce stress anxiety.

In another earlier experiment, Brennan et al. (2000) studied the effect of music in reducing stress anxiety by bringing music into the workplace. Their experiment studied the effect music has on self-reported stress levels and the immune system. The researchers obtained baseline levels of self-reported stress and measured the presence of salivary immunoglobin A (S-IgA) in the blood of the participants. S-IgA usually is an indicator of immune system activity. This pretest and posttest experimental design was designed in such a way that after playing smooth jazz in the participants' high-stress newsroom workplace, they measured the participants' levels of self-reported stress and the levels of S-IgA in the blood. Music was shown to have a reduction on the participants' perceived level of stress anxiety. Researchers have continued to show the effect of music on human health. Brownley et al. (1995) carried out a research on the influence of music on physiological and affective exercise responses in 8 trained and 8 untrained runners under three music conditions ("no", "sedative", and "fast") during low, moderate, and high intensity exercise. Repeated results of this study revealed increased respiratory frequency (greater excitation) during fast music as compared to the no music and sedative music conditions.

In a study on music and labor pain, Geden et al., (1989) carried out two studies to examine the effects of music on analogue labor pain using volunteer nulliparous subjects who were randomly assigned to treatment groups. Assessments of the treatments were made in a 1-hour session involving twenty 80-second exposures to a laboratory pain stimulus patterned to resemble labor contractions. In the first experiment, it was hypothesized that subjects listening to easy-listening music would report lower pain ratings and cardiovascular responses than subjects listening to rock music, self-selected music, or a dissertation (placebo-attention) and subjects in a no-treatment control group. Significant group effects were found between each listening and all other groups. Subjects spontaneously reported using imagery as a pain reduction technique. In the second study a combination of music and imagery was examined by randomly assigning subjects to one of five groups: self-generated imagery with music (SIM), guided imagery with music (GIM), self-generated imagery without music (SI), guided imagery without music (GI), or no-treatment control. Again, significant group effects were obtained between with music and without music for all groups.

In another study on the effect of music on stress anxiety levels, Migneault et al. (2004) examined the effect of music on the neurohormonal stress response to surgery under general anesthesia. Their study found that several pharmacological interventions reduce preoperative stress hormone release during surgery under general anesthesia. They also studied listening to music and therapeutic suggestions where they found a positive effect on postoperative recovery and the need for analgesia. In this study, they evaluated the effect of listening to music under general anesthesia on the neurohormonal response to surgical stress as measured by appropriate medical procedures. Music therefore had an effect even under anesthesia.

In their research where they attempted to the answer the question: “Does Singing Promote Well-Being?”, Grape et al., (2004) carried out an empirical study of professional and amateur singers during a singing lesson. This study explored the possible beneficial effects of singing on an individual’s well-being during a singing lesson. Eight amateur (2m, 6f, age 28-53 yrs) and eight professional (4m, 4f, age 26-49 yrs) singers who had been attending singing lessons for at least six months were included. Five visual analogue scales (VAS, sad-joyful, anxious-calm, worried-elated, listless-energetic, and tense-relaxed) were scored before and after the lesson. In addition, a semi-structured interview was performed. Heart rate variability analyses showed significant changes over time in the two groups for total power, and low and high frequency power. Power increased during singing in professionals, whereas there were no changes in amateurs. This indicates an ability to retain more "heart-brain connection." i.e., more cardio-physiological fitness for singing in professional singers, compared to amateur singers. In another study on anxiety, Yilmaz et al. (2003) carried out a study to find out if music decreases anxiety and provides sedation in a procedure known as extracorporeal shock wave lithotripsy (ESWL). In this procedure, sound waves to break a kidney stone into small pieces that can more easily pass into the bladder. This study revealed that listening to music by patients during (ESWL) treatment is a feasible and convenient alternative to sedatives and anxiolytics.

Winter et al. (1994) observed that it is well known that many patients become stressed and anxious prior to and after surgery and that one means of helping reduce anxiety in patients is to incorporate music in the Surgical Holding Area.

In their study, they divided their subjects into two groups: one group of patients listened to music while a second group did not. The researchers observed that patients who listened to music while in the Surgical Holding Area had significantly less stress and anxiety than did those who did not listen to music. Both groups spent similar lengths of time in the Surgical Holding Area. Winter et al. concluded that patients preferred the musical intervention since they experienced less anxiety.

In another study, researchers at the Bryan Memorial Hospital, Lincoln, investigated the influence of music therapy on mood and anxiety of patients undergoing heart surgery. Ninety-six patients who underwent elective, heart bypass surgery at the cardiovascular intensive care and progressive care units of a Midwestern community hospital participated in the study. Data relating to anxiety and mood was obtained through blood pressure and heart rate using the Spielberger's state-trait anxiety inventory (STAI). The study revealed that the patients' mood ratings showed significant improvement in mood among those patients who were in the "music therapy" group after the second intervention. There was also significant main effects over time for heart rate and systolic and diastolic blood pressure in the music therapy group , which indicated a generalized physiologic relaxation response. Reduced anxiety and improved mood were observed in all three groups, and the researchers noted that all of the interventions demonstrated a generalized relaxation.

In another study, Augustin et al. examined the role that music therapy might play in a postoperative setting for ambulatory patients. In the Day Surgery Unit of St. Mary's Hospital, Mequon, Wisconsin, they investigated Forty two ambulatory surgery patients where they assigned them to either an experimental group to receive music therapy along with the standard preoperative instructions or a control group to receive the standard preoperative instruction alone. The results revealed that the patients in the experimental group showed significantly lower heart rates compared to the patients in the control group. The experimental group also showed greater improvements in blood pressure and respiration rate. The researchers concluded that music therapy offers demonstrable benefits for ambulatory surgery patients and recommend that the patients be offered music as an effective option to help alleviate postoperative anxiety.

The effect of music in the alleviation of physiological conditions such as anxiety is a historical phenomenon. This practice is evidenced by several mythological reflections, which have explained the origins of music in healing. One belief derives the use of music in healing from the Old Testament in the Bible (circ. 1445 B.C.) where David played his harp to Saul for curative purpose (Benenzon, 1997; Moreno, 1988; Darrow, Gibbons & Heller, 1985). Some attribute the rise of music in healing to the life in Stone Age, which is saturated with the practice of using music for health. The Stone Age cultures did not view music as a self-contained experience apart from life.1 The connection between music and healing has come a long way from its primitive stage; but this connection has been retained with the human existence. In this regard, we may speak of music as a universal agent that transcends all time and space, barriers withstanding in alleviating suffering such as that caused by pain and anxiety. Its use as a healing agent has been shaped over thousands of years through human experience.

METHOD

Participants

With a target sample size of 100, twenty five undergraduate music students ranging in age from 18 to 24 years participated in the study. All subjects were either majoring in or studying music as one of their minor subjects. The subjects practiced music in the practice rooms in a university in the USA.. The selection process was based on opportunity sampling. Thus there was no random sampling but those who wished to participate in the study were recruited and given the Sate Trait Anxiety Inventory questionnaire to fill 10 minutes before practice and immediately after practice. Criteria for subject participation included (a) being able to practice music on any instrument for at least half an hour in order to participate in the study. (b) The subjects had to be at least in their intermediate level of musicianship. (c) The subjects had to be at lease 17-21 years of age. (d) The subject's primary language must be English (as only the English version of the STAIC scale was used). Prior to participation in the study, subjects were informed of the study's purposes. If the subjects agreed to participate, they reviewed and signed a consensus form.

Materials

Informed Consent forms were used which informed the participants of the purpose of the study and explained that the participants would be filling out two surveys. The tool used to measure anxiety was The State-Trait Anxiety Inventory (STAI), which was invented, by Charles Spielberger, Richard L. Gorusch, and Robert E. Lushene in 1957. Forms C-1 and C-2 of the State Trait Anxiety Inventory (STAI) were used to measure the participants' current level of stress anxiety on two occasions. According to Charles et al., state anxiety reflects a "transitory emotional state or condition of the human organism that is characterized by subjective, consciously perceived feelings of tension and apprehension, and heightened autonomic nervous system activity." State anxiety may fluctuate over time and can vary in intensity. In contrast, trait anxiety denotes "relatively stable individual differences in anxiety proneness . . ." and refers to a general tendency to respond with anxiety to perceived threats in the environment.

Design and Procedure

The study used a single-variable pre-test/ post-test within groups design. The independent variable was music and practice time while the independent variable was the anxiety. Subjects were required to do fill out the STAIC forms C-1 and C-2 at least 10 minutes before their practice. They had to practice at least for 30 minutes without interruption. After their practice, the subjects had to complete the STAIC forms C-1 and C-2 one more time. Every subject did this only once. The type of methodology used here was within subjects. All the subjects participated in the treatment group. Before beginning the experiment, informed consent was obtained from every participant. Once all of the participants arrived, the piano practice room door was shut and the STAI was passed out along with standard #2 pencils. After writing their code number in the upper right hand corner, they were told to read the instructions and fill out the STAI according to how they felt at that moment. At this point, no music had been played. The participants were informed that they would have 7 minutes to complete the self-report. After they finished, the surveys were collected and then the participant was left to practice for half an hour. In addition, they were informed that they had participated in an experiment studying the effects of music practice on stress anxiety.

They were also informed that all of their information was anonymous and that the answers they provide would be compiled statistically. Finally, after the practice, the participants carried out another set of self-repot on the C-1 and C-2 questionnaires.

RESULTS

I hypothesized that the posttest measures of stress anxiety should not differ. The results obtained did not support this hypothesis. (pretest State = 34.8, posttest = 33.44). I also hypothesized that pretest and posttest measures of Trait should not differ. This hypothesis was supported by the results (pretest Trait = 34.48, posttest Trait =34.64). Furthermore, I hypothesized that post measures of stress anxiety for State would differ significantly with the sexes with f being the least stressed and the m group being the most stressed. This hypothesis was not supported by the results obtained (male M= 33.2, female M= 31.2)

However, there was a relationship between the pre-test and the post-test STAI scores (r =0.903)

DISCUSSION

The researcher hypothesized that same levels of stress anxiety would be reported after the music practice in every participant. It was also hypothesized that females would not show any significant differences in the pretest and posttest anxiety levels from those of males.

The present data were not found to be significant and the null hypothesis cannot be rejected. The mean scores went down for both sexes on the posttest of the STAI but it did not reach significance. Even though there were very minimal controls to this study, physiological measures of heart rate, respiration rate, blood pressure, skin temperature, EEG, and neurohormone levels would all have been beneficial and would provide more comprehensive assessment. In this study, one respondent indicated that he often felt better after his music practice. Another respondent said that he normally practices for long hours and that the longer he played the better he felt.

Some possible reasons why the data is not congruent with past research could be the length of time that the participants were exposed to music and the type of music they actually played. The other reason could be that the participants played music only for a short period of time. In this experiment, the participants were exposed to music for a short period of time, about thirty minutes, whereas in much of the past research, participants were exposed to the music for longer periods of time (Hammer, 1996), and in some cases an entire day (Brennan et al., 2000). Participants were exposed to music for only a short period of time in this study in order to increase the external validity of the study. People often cannot practice music all day, but maybe only while they are driving or working out. In an effort to increase the external validity, the internal validity was compromised and that could have contributed to the data not being significant.

Another possible reason could be the use of the STAI. Some of the past research used more objective measurements such as salivary recordings and by measuring nervous system activity (Brennan et al., 2000; McCraty et al., 1996). Because the participants were administered two similar surveys within a short period of time, their answers on Form C-2 might have been influenced by their answers on Form C-1, an idea supported by the correlation between the pre-test and post-test scores. The high positive correlation (r= .903) between participants' answers on the two surveys indicates that their answers on the first survey would be the best predictor of their answers on the second survey, rather than the type of music they practiced. The study did discover some significant relationship between the sexes and how their anxieties fluctuate with music practice. General Linear Model indicated that the State anxiety overall decreased across time. There was a trend for women to show a larger decrease than males but it does not reach significance suggesting that sex might not have as much of an effect as was originally hypothesized. Because the participants were only exposed to the music for a short time, the results could be indicative of the fact that participants must be exposed to music for a much longer time for the music to significantly affect stress anxiety levels. Therefore, practicing music will not significantly lower stress anxiety levels if people only practice for short periods of time. Also, music therapy was not incorporated into this study. Music therapy, such as the guided imagery used in Hammer's (1996) experiment or the positively induced emotional state used in McCraty et al's (1996) experiment, compels the participant to become actively involved in reducing their stress levels while playing music. This study could have implications for people who passively practice specific types of music intended to relieve stress, for short amounts of time and think that they are lowering their stress levels. The results of this study indicate that this belief is not valid. People may need to practice music throughout the day and incorporate some type of music therapy, such as guided imagery, into their music experience to gain any of the possible benefits.

Because this study did not discover a significant relationship between music practice and stress anxiety, it could be viewed as weakening the prevailing view that music does indeed affect stress anxiety. But, rather than negate the bulk of the past research (Hammer, 1996; Brennan et al., 2000; Wiesenthal et al., 2000; Standing et al., 1980) that does support the hypothesis that music does affect stress anxiety, the results of this study could have greater implications for the perceived strength of the relationship between stress and music practice. Since music was only practiced for a short period of time, the relationship between music practice and stress anxiety would have to be strong for the results to be significant. A relationship could exist, but it might be too weak to have been demonstrated in this study.

For future studies, researchers should expose the participants to music practice for a longer period of time so that the music could have a greater effect. Also, using biological methods to measure stress anxiety such as physiological measures of heart rate, respiration rate, blood pressure, skin temperature, EEG, and neurohormone levels are preferable if it possible. These methods are more expensive, but provide more objective and reliable data.

It would be interesting to do further studies on the effects of forced music practice on stress anxiety. Future studies could also focus on the effects of music practice on stress anxiety and the physical health of the participant and also focus on the strength of the relationship. The results of future studies could be used to implement simple and inexpensive ways to alleviate stress anxiety in the general population by providing music practice rooms at in the hospitals, schools, work places, congress and the factories, etc.

References

Akombo, D.O. (2013). *Effects of Community African Drumming on Generalised Anxiety in Adolescents*. Approaches: Music Therapy & Special Music Education 5 (1) :25-33.

Aitken J., Wilson S., Coury D., & Moursi AM. (2002). *The effect of music distraction on pain, anxiety and behavior in pediatric dental patients*. Pediatric Dentistry 24 (2)114-8,

Aldridge, D. (1993). *Artists as psychotherapists.* Arts in Psychotherapy, 20, 3, 199- 200. Allen K., Golden L.H. Izzo JL Jr.

Ching MI. Forrest A. Niles CR. et al., (2001). *Normalization of hypertensive responses during ambulatory surgical stress by perioperative music.* Psychosomatic Medicine. 63(3)487-92

Augustin P.& Hains AA. (1996). *Effect of music on ambulatory surgery patients' preoperative anxiety.* AORN Journal 63(4) 750, 753-758

Baines, S. (2000). *A consumer-directed and partnered community mental health music therapy program: Program development and evaluation.* Canadian Journal of Music Therapy, VII (1) 51-70.

Barnason S., Zimmerman L., Nieveen J. (1995). *The effects of music interventions on anxiety in the patient after coronary artery bypass grafting*. Heart Lung 24 (2) p124-32

Berger, D. S. (1999). *Toward the Zen of performance: Music improvisation therapy for the development of self-confidence in the performer.* MOB Music, Inc., St. Louis, MO.

Bettermann H., von Bonin D., Fruhwirth M., Cysarz D., & Moser M. (2002). *Effects of speech therapy with poetry on heart rate rhythmicity and cardiorespiratory coordination.* Herdecke, Germany: Department of Clinical Research, Gemeinschaftskrankenhaus, 58313,. International Journal of Cardiology 84(l) 77-88.

Bell, T.P. & Akombo, D.O. (In Print). *Long-term music-listening's effects on blood pressure, heart rate, anxiety and depression*. Journal of Alternative Medicine Research, 2017, 9(1).

Blood A., Zatorre R.(2001). *Intensely pleasurable responses to music correlate with activity in brain regions implicated in reward and emotion.* Montreal, QC, Canada: Montreal Neurological Institute, McGill University,. Proceedings of the National Academy of Sciences of the United States of America. 98 (20) 11818-23.

Bonnet M.,& Arand D. (2000). *The impact of music upon sleep tendency as measured by the multiple sleep latency test and maintenance of wakefulness test. Dayton: Dayton Department of Veterans Affairs Medical Center.* Physiology & Behavior. 71 (5) :485-92.

Brennan, F.B., & Charnetski, C.J. (2000). *Stress and immune system function in a newspaper's newsroom.* Psychological Reports, 87, 218-222.

Broscious S. (1999). *Music: an intervention for pain during chest tube removal after open heart surgery.* Newport News, Va: Christopher Newport University. American Journal of Critical Care 8(6) 410-5.
Browning C. (2000). Using music during childbirth. Ontario, Canada: Wilfrid Laurier University in Waterloo. Birth. 27(4) 272-6,

Brownley K., McMurray R.,& Hackney A. (1995) . *Effects of music on physiological and affective responses to graded treadmill exercise in trained and untrained runners.* Chapel Hill: International Journal of Psychophysiology 19 (3) 193-201.

Campbell, M. (1982). *Music therapy students in concert.* British Journal of Music Therapy, 13 (1) 8-15.

Cottrell, A. (2001). *Healing Musical Instruments.* Retrieved September 15, 2016 from the World Wide Web at: http://www.healingmusic.org/Main/catalog/Drums.htm

Cruise C., Chung F., Yogendran S.,& Little D. (1997) . *Music increases satisfaction in elderly outpatients undergoing cataract surgery.* Ontario, Canada: Department of Anaesthesia, University of Toronto. Canadian Journal of Anaesthesia 44(l) 43-8.

Cudney S., Butler M., Weinert C.,& Sullivan T. (2002). *Ten rural women living with fibromyalgia tell it like it is.* Bozeman, Montana: College of Nursing, Montana State University-Bozeman. Holistic Nursing Practice. 16(3) 35-45.

Davis, W. (1987). *Music therapy in Nineteenth Century America.* Journal of Music Therapy, XXIV (2) 76-87.

__________. (1989). *Music therapy in Victorian England: Frederick Kill Hartford and the Guild of St. Cecilia.* Music Therapy Perspectives, Vol. 7, pp. 17-22.

Diaz de Chumachiero, C.L. (1990). *Dynamic implications of opera training in psychotherapy.* Arts in Psychotherapy, 17, (1). 19-28.

Elliot, D. (1992). *Rethinking music teacher education.* Journal of Music Teachers Education, Fall 1992, pp. 6-15.

Evers S., Dannert J., Rodding D., Rotter G.,& Ringelstein E. (1999). *The cerebral haemodynamics of music perception. A transcranial Doppler sonography study.* Munster, Germany: Department of Neurology, University of Munster. Brain. 122 (Pt l) 75-85.

Erdtsieck, J. (1997) *Pepo As An Inner Healing Force: Practices of a Female Spiritual Healer in Tanzania.* Amsterdam: Royal Tropical Institute.

Fenwick, A. (1971). *From silence to speech: A case-study of a schizophrenic patient.* British Journal of Music Therapy. pp. 1-6.

Friedson, S. (1998) . *Tumbuka Healing.* In The Garland Encyclopedia of World Music Vol. 1, R.M. Stone (Ed). New York: Garland Publishers.

Geden E., Lower M., Beattie S.,& Beck N. (1989). *Effects of music and imagery on physiologic and self-report of analogued labor pain.* Columbia: Nursing Research. 38(1) 37-41, 1989.

Gerra G., Zaimovic A., Franchini D., Palladino M., Giucastro G., Reali N., et al.,(1998). *Neuroendocrine responses of healthy volunteers to 'techno- music': relationships with personality traits and emotional state.* Centre Studi Farmacotossicodipendenze, Az. USL di Parma, Italy. International Journal of Psychophysiology 28 (1) 99-111.

Good, Charles M (1987) *Ethnomedical systems in Africa: patterns of traditional medicine in rural and urban Kenya.* New York: Guilford Press.

Grape C., Sandgren M., Hansson L., Ericson M.,& Theorell T. (2003). *Does singing promote well-being?: An empirical study of professional and amateur singers during a singing lesson.* Stockholm: Sweden. Integrative Physiological & Behavioral Science 38(l) 65-74.

Hammer, S.E. (1996). *The effects of guided imagery through music on state and trait anxiety.* Journal of Music Therapy, 33, 47-70.

Harris L. (1996). *Horn playing and blood pressure*: Lancet. 1996 Nov 30;348(9040):1528; PMID: Lancet. 348 (9033) 1042, 1996.

Harper, P. (1969). *Dance In Nigeria.* In Ethnomusicology, 9 (1), 280-295.

Heiser R., Chiles K., Fudge M., Gray S. (1997). *The use of music during the immediate postoperative recovery period.* Lexington: OR Services, University of Kentucky Hospital. AORN Journal. 65(4) 777-8, 781-5.

HealthyPlace (n.n). *Depression, Anxiety, and Other Mental Illnesses Are Inter-related.* Retrieved September 14th 2016 from the World Wide Web at http://www.healthyplace.com/communities/depression/related/depression_anxiety_2.asp

Heller, G. (1987). *Idea, initiatives, and implementations: Music therapy in America, 1789-1848.* Journal of Music Therapy, 24(1) 35-46.

Hollander, L. (1988). *Music in the creative process and the path of enlightenment.* Music Therapy, 7 (1) 8-14.

Isern, B. (1964). *Music in special education.* Journal of Music Therapy, Vol. 4, pp. 135-142.

Horden, P. (2000). *Music as Medicine: The History of Music Therapy since Antiquity.* Hants: Ashgate Publishing House.

Iwanaga M. (1995). *Relationship between heart rate and preference for tempo of music.* Higashi- hiroshima, Japan: Department of Behavioral Sciences, Faculty of Integrated Arts and Sciences, Hiroshima University. Perceptual & Motor Skills 81(2):435-40.

Janzen, John M (1992). *Ngoma : Discourses of Healing in Central and Southern Africa.* Berkeley : University of California Press.

___________ (1978). *The Comparative Study of Medical System as Changing Social Systems* In Social Science and Medicine (12) 2B: 121-129

Josepha, M. (1964). *The therapeutic values of instrumental performance for severely handicapped children*. Journal of Music Therapy, September, pp. 73 –79.

Konlaan B., Bjorby N., Bygren L., Weissglas G., Karlsson L., Widmark M. (2000). *Attendance at cultural events and physical exercise and health: a randomized controlled study.* Umea, Sweden: Department of Social Medicine, University of Umea. Public Health. 114(5) 316-9.

Landolt M., Marti D., Widmer J., Meuli M. (2002). *Does cartoon movie distraction decrease burned children's pain behavior?* .Zurich, Switzerland: Pediatric Burn Center, Department of Surgery, University Children's Hospital Zurich, Switzerland. Journal of Burn Care & Rehabilitation. 23(l) 61-5.

Linden, S. (1997). *A festival of light: A high school healing arts event celebrating diversity of the school community*. Arts in Psychotherapy, 24 (3) 255-259.

Machlis, Joseph (1963) . *The Enjoyment of Music.* New York: W.W. Norton & Company.

Lundgren I., Dahlberg K. (2002). *Midwives' experience of the encounter with women and their pain during childbirth.* Boras, Sweden: Boras University College of Health Sciences. idwifery 18(2) 155-64.

Magill L. (2001) *The use of music therapy to address the suffering in advanced cancer pain.* New York: Integrative Medicine Service, Memorial Sloan-Kettering Cancer Center, New York University, , New York, USA. Journal of Palliative Care. 17(3):167-72, 2001 Autumn.

Martin R. (2001). *Humor, laughter, and physical health: methodological issues and research findings.* London, Ontario: Department of Psychology, University of Western Ontario. Psychological Bulletin. 127(4) 504-19.

Matz A., Brown S., (1998). *Humor and pain management. A review of current literature.* Wilmington, North Carolina: Cape Fear Community College. Journal of Holistic Nursing. 16(l) 68-75.

Migneault B., Girard F., Albert C., Chouinard P., Boudreault D., Provencher D., et al., (2004). *The effect of music on the neurohormonal stress response to surgery under general anesthesia.* Montreal: Anesthesia & Analgesia. 98(2) 527-32.

Miller, J. (1998). *The contributions of Wayne Rupenthal to the field of music therapy.* Journal of Music Therapy, Vol. 35, No. 2, pp. 105-118.

Montello, L., Coons, E.E., & Kantor, J. (1990). *The use of group music therapy as a treatment for musical performance stress.* Medical Problems in Performance Art, Vol. 5, pp. 49-57.

Montello, L. (1989). *Utilizing music therapy as a mode of treatment for the performance stress of professional musicians.* Doctoral dissertation, New York University.

Newman, F. & Holzman, L. (1997). *The end of knowing.* London. Orne, M. (1959). Journal of Social Psychology, 58, 277

Petrie A. (1960). *Individuality in Pain and Suffering.* Chicago: University of Chicago Press.

Patrick G. (1999). *The effects of vibroacoustic music on symptom reduction.* IEEE Engineering in Medicine & Biology Magazine. 18 (2) 97-100.

Pickett, M. (1976) . *Music therapy in the community.* British Journal of Music Therapy, pp. 31-37.

Ragland & Aprey (1974). *Community music therapy with adolescents*. Journal of Music Therapy, Vol. 11, No. 3, pp. 147-155.

Reubart, D. (1985). *Anxiety And Musical Performance.* New York: Da Capo Press.

Rupenthal, W. (1965). *Scribbling in music therapy.* Journal of Music Therapy, Vol. 2, No. 1, pp. 8-10.

Rutledge.Oddy, N. (2001). *Convergencies: Possibility for therapeutic intervention in a large scale community performance*. Canadian journal of Music Therapy, VIII (1) 48-63.

Salmon, S. (1981). *Music therapy with maladjusted children.* British Journal of Music Therapy, Vol. 12, No. 3, pp. 4-11.

Schmidt L., Trainor L., Santesso D. (2003). *Development of frontal electroencephalogram (EEG) and heart rate (ECG) responses to affective musical stimuli during the first 12 months of post-natal life.* Ontario: Canada. Brain & Cognition 52(l) 27-32.

Schneider N., Schedlowski M., Schurmeyer T., Becker H. (2001). *Stress reduction through music in patients undergoing cerebral angiography*. Hannover: Department of Neuroradiology, Hannover Medical School, Germany. Neuroradiology. 43(6 :472-6.)

Standing, L. & Stace, G. (1980). *The effects of environmental noise on anxiety level.* The Journal of General Psychology, 103, 263-272.

Stige, B. (2000). *The voice and warmth of a cello - introduction to Juliette Alvin's article.* Nordic Journal of Music Therapy,. Vol. 9, No. 1, pp. 46-49.

Szmedra L., Bacharach D. (1998). *Effect of music on perceived exertion, plasma lactate, norepinephrine and cardiovascul hemodynamics during treadmill running.* Bloomsburg: Department of Kinesiology, Bloomsburg University. International Journal of Sports Medicine. 19(l) 32-7.

Taylor, D. (1981). *Music in general hospital treatment from 1900 to 1950*. Journal of Music Therapy, Vol. 18, No. 2, pp. 62-73.

Tsuchiya M., Asada A., Ryo K., Noda K., Hashino T., Sato Y., Sato E., Inoue M. (2003). *Relaxing intraoperative natural sound blunts haemodynamic change at the emergence from propofol general anaesthesia and increases the acceptability of anaesthesia to the patient.* Acta Anaesthesiologica Scandinavica. 47 (8) 939-43.

Umemura M., Honda K. (1998). *Influence of music on heart rate variability and comfort—a consideration through comparison of music and noise.* Tokyo: Department of Management Science, Science University of Tokyo, Shinjuku-ku. Journal of Human Ergology 27(1-2) 30-8.

Wang S., Kulkarni L., Dolev J., Kain Z. (2002). *Music and preoperative anxiety: a randomized, controlled study*. New Haven: Department of Anesthesiology, Yale University School of Medicine. Anesthesia & Analgesia 94(6) 1489-94.

Winter M., Paskin S., Baker T. (1994*). Music reduces stress and anxiety of patients in the surgical holding area.* J Post Anesth Nurs, 9 (6) 340-3

Yamamoto T., Ohkuwa T., Itoh H., Kitoh M., Terasawa J., Tsuda T., Kitagawa S., Sato Y. (2003). *Effects of pre-exercise listening to slow and fast rhythm music on supramaximal cycle performance and selected metabolic variables.* Archives of Physiology & Biochemistry. Ill(3) 211-4.

Yilmaz E., Ozcan S., Basar M., Basar H., Batislam E., Ferhat M. (2003). *Music decreases anxiety and provides sedation in extracorporeal shock wave lithotripsy.* Kirikkale: Department of Urology, Kirikkale University Medical Faculty. Urology 61(2):282-6.

Contributors

David O. Akombo, Ph.D. is Associate Professor, Director of Music Education and Coordinator of Graduate Studies at Jackson State University. Before joining Jackson State University, Dr. Akombo was a Diversity Fellow at Weber State University in Utah. Intrigued by the peculiarities of music including the interdisciplinary studies of music in medical and psychological practices, Dr. Akombo has studied and researched the effects of music on children and young adults both in schools and in hospitals. As scholar and performer, Dr. Akombo balances his research interests amicably as a music educator, ethnomusicologist, composer and singer/drummer, having worked in Africa and Southeast Asia where he studied with Balinese artists. His books, Music and Healing Across Cultures (Ames, Iowa: Culicidae Press, 2006) and Music and Medicine: Connections Found (Seaburn, 2009) unfold the mechanics of the relationships between music, culture, physiological and spiritual states in people from the anthropological and epistemological inquiry regarding music healing and the cosmos. These books show the power of music and its ability to promote mind/body coordination. Dr. Akombo has worked jointly with the University of Florida's Center for Arts in Healthcare Research and Education (CAHRE) with which he collaborated to help establish the Arts in Medicine (AIM) program in Nairobi, Kenya. Located at Mater Hospital and funded by the State of Florida Center for Cultural Affairs in Tallahassee, this is the first known Arts in Medicine program to be established in Africa. Dr. Akombo continues to be active in music education research, world music, interdisciplinary scholarship in music and he has spoken and conducted clinics in several countries. He is a member of the Music Educators National Conference, the Utah Music Educators Association, Society for Ethnomusicology, and Society for the Arts in Healthcare. Dr. Akombo is a co-founder and a member of the Executive Board of the Interdisciplinary Society for Quantitative Research in Music and Medicine. dakombo@isqrmm.com

Mary Elizabeth Campbell, CAGO, MAMT-BC 'Mib' has her BA from Wilson College in Chambersburg in 1980, where she majored in Organ Performance and Religion Studies and minored in Psychology and Fine Arts (always one to have trouble making choices...) After Wilson she studied organ with Claribel Thomson in Ardmore, PA, completed an MA in Social Gerontology at the University of Pennsylvania in 1984. She worked in the adult day care field from the 1980's until 1994, serving as director for the Devon Manor Adult Day Center and subsequently the Tel Hai Adult Day Center. She left full time work when her first husband died, and at that time earned her second MA in Music Therapy at Immaculata in 1997. She currently does part time group therapy in the intensive mental health outpatient department of Creative Health in Pottstown. She has been affiliated with 'A Touch of Music,' a music therapy provider in Pottstown, since 1995, and consults at several agencies for them, running music therapy groups with populations ranging from older dementia and stroke patients to developmentally disabled adults with a wide range of abilities. She has supervised graduate and undergraduate interns and is now serving in that capacity for Immaculata University. In 2007 and 2008 she was trained in EMDR by Dr. Laurel Parnell, including advanced level clinician training, and since that time has seen some private clients for EMDR. She also teaches part time for the Community Health Education Department at Phoenixville Hospital. sunflowerangels@verizon.net

D. Weldon Cochren, M.Div. is a music producer, songwriter, and sound designer. Mr. Cochren has worked with many recording artists and recording companies in the United States and Canada. He has studied Pentecostalism and its African/African American roots for a number of years, and has been involved in Pentecostal ministry over 37 years. Mr. Cochren has ministered and taught domestically and in Canada, South Africa, Zimbabwe, Botswana, and Ghana. Mr. Cochren's extensive experience in religion, spirituality and music are the major factors that caused him to research and investigate the possible connectivity to human wholeness. Mr. Cochren earned a Master of Divinity from Drew University in Madison, New Jersey and was granted numerous scholarships and awards. He focused much of his studies in the areas of "African-Based Religions of the Caribbean and Latin America," "Religion and Culture," "Music of the World's Religions" and "Music of the Emerging Church." Mr. Cochren has a Bachelor of Science degree in Arts Management from Empire State College, State University of New York. wcochren@hotmail.com

Laura Dunbar, Ph.D. has taught 10 years in the public schools beginning her career as a band director teaching instrumental music to grades 5-12 at Elmwood Local Schools in Ohio. When her family relocated to Arizona, she accepted a K-8 position in Sedona, Arizona where she taught general, instrumental, vocal, and steel drum music for seven years. Laura was a graduate assistant at the University of Arizona where she completed a Ph.D. in Music Education with a minor in Educational Psychology. Her primary research interests include the effects of music and music training on those with attention deficits. She has presented at local, state, and national conferences and in-service workshops. Laura is currently Assistant Professor of Music Education-Elementary Specialist and Music Education Coordinator at the University of Wisconsin-Eau Claire. DunbarLL@uwec.edu

Steve Jackowicz, Ph.D. began his study of East Asia as an undergraduate at Harvard University. He then studied traditional East Asian Medicine at the Korea Modern School of Acupuncture and the Hand and Foot Acupuncture Institute in Seoul. During that time he further apprenticed in traditional physical manipulation and therapeutic breathing techniques. He then pursued a Master's Degree at the New England School of Acupuncture. Dr. Jackowicz became fascinated with the divergence of Chinese, Japanese, Korean, and Vietnamese medicine and studied for a Ph.D. at Boston University examining the historical evolution of East Asian Medicine. His research has taken him to China and Japan. Dr. Jackowicz has taught at Tufts University, Boston University, New England School of Acupuncture, and Pacific College of Oriental Medicine. He has published on East Asian Medicine, Daoism, and history. He currently serves as faculty at the University of Bridgeport. stevejackowicz@gmail.com

Roy Kennedy, Ph.D., MT-BC, is an associate professor of music therapy at the University of Georgia, USA. While previously providing music therapy services to various populations, his primary research interests include guitar pedagogy for music therapists, music therapy for English as Second Language learners, adults with substance abuse problems, and skill generalization of students with developmental disabilities. Dr. Kennedy has presented his work in the USA, Brazil, South Korea, and Costa Rica. He continues having an interest in the international development of the profession. rkennedy@uga.edu

Kristie Lipford, Ph.D. is a NIH Disparities Research Trainee and a junior faculty member at Jackson State University where she divides her time between teaching and research. In her teaching role, Dr. Lipford oversees and plans curriculum for the Research Methods course in the Bachelor of Social Work Program. In her research role, she broadly examines health behaviors. She is currently completing a manuscript that examines the effect of psycho-social stressors on physical activity, dietary habits, and sleep patterns using Jackson Heart Study data. Her most recent focus though is integrative medicine. At the present, she is preparing a NIH Career Development Award to undergo a review with the National Center for Complementary and Integrative Health. Dr. Lipford is interested in analyzing locally, nationally, and globally available datasets to examine complementary and alternative medicine use among ethnic groups and minority populations. kristie.j.lipford@jsums.edu

Eric Miller, Ph.D. MT-BC BCB biofeedback therapist and board-certified music therapist is author of Bio-guided Music Therapy (2011) Jessica Kingsley Publishers. He serves as adjunct faculty at Immaculata University, Director of the Ott Lab for Music & Health Coordinator at Montclair University, and has lectured at the Naropa Institute and Bryn Mawr College among other institutions. Recent international presentations include sessions at the World Music Therapy Congress in Seoul, S. Korea, Hsien Chuan University in Tapei, Taiwan, and workshops in Kiental, Switzerland, Cogolin, France and Torino, Italy. Dr. Miller serves as Executive Director of nonprofits Music for People, Expressive Therapy Concepts and founded the Biofeedback Network. In his spare time, Eric chairs the Pennsylvania Badminton Association. Eric collaborated with Grammy winning cellist, David Darling on the instrumental CD, *Jazzgrass*. Dr. Miller may be reached at: miller@biofeedback.net

Kaleb A. Santana Ramos was born and raised in the island of Puerto Rico. He holds both a bachelor degree in percussion and music education. He is currently working on a master's degree in music education from the Conservatory of Music of Puerto Rico. Mr. Santana Ramos is a certified music teacher in Puerto Rico. He specializes in classical and popular percussion individual and group lessons. As a private drum set and percussion teacher he has worked with students with Cerebral Palsy, DNA disorder, and Speech Apraxia. In October 2013, he presented his thesis research entitled The Drum Set and Cerebral Palsy: Motor and Affective Psychological Development at the Music Education Forum in the Conservatory of Music of Puerto Rico. In March 2015, he also presented this same research study at the XIII Puertorrican Research Congress in Education at the University of Puerto Rico. KalebSantana@gmail.com

Lynn Saltiel is a Licensed Clinical Social Worker and a Certified EMDR Therapist currently in private practice. She also is a Certified Music Improvisation Teacher from Music For People. Lynn holds her Masters of Science in Social Work from Columbia University and has been an Instructor for the National Association of Social Workers-New Jersey Chapter (NASW-NJ), since 2002. She divides her practice time doing psychotherapy, training, and supervision. She presents for many Fortune 500 companies providing employee health and wellness seminars. Lynn developed and facilitated community programs for The Women's Project and Women Rising, servicing and empowering at risk women utilizing music and the creative arts. She is a musician trained in the use of facilitation and music improvisation techniques. Lynn is a member of the Open Music Ensemble in NYC and the Dance Improv Ensemble in Princeton, NJ. She has presented at Flute Haven Native Flute School, NASW-NJ, NYSCSW, New Jersey City University and the Globe Sound Healing Conference. jlynnlcsw@gmail.com

Emma Rodríguez Suárez, Ph.D. was born and raised in the Canary Islands, Spain. She holds a B.M. in M.Ed. and a M.M.Ed. from the Hartt School of Music, University of Hartford. She also holds a Ph.D. in M.Ed. from the University of Toronto. Dr. Rodríguez Suárez is a teacher trainer and holds a Level III Orff-Schulwerk Teacher Training certificate and a Kodály Certificate from the Kodály Musical Training Institute. She was awarded an Artist-Teacher Certificate by the Association for Choral Music Education and she also holds a Creating Artistry Conductors Workshop Level III certificate. Dr. Rodríguez Suárez is the author of Canciones de mi Tierra Española: Islas Canarias/Songs of my Spanish Land: Canary Islands and has published numerous articles, and presented workshops both nationally and internationally. Dr. Rodríguez Suárez is an Assistant Professor of music education at Syracuse University and a visiting lecturer at the Conservatory of Music of Puerto Rico. singingcanaries@gmail.com

Naejeong Suh, Ed.D was born and raised in Seoul, South Korea. She has taught general and instrumental music in elementary and middle school in South Korea. She also has worked as a concert violist, performing numerous recitals and participating in band and orchestra in Seoul, Korea, and the U.S. She holds a Bachelor's Degree in violin performance and Master's Degree in Music education from the Kyung Hee University of Korea. She also holds an Ed.D in Music Education from University of Georgia. Her research interests include the effects of music on the quality life of the healthy elderly people. naejeong.suh@gmail.com

Wendy Thompson, Ph.D. is an Associate Professor of the Graduate Social Work Program at Jackson State University. She previously served as an Associate Professor of Social Work at Andrews University for three years where she taught research methods, human behavior, critical thinking, dealing with difficult people, families and children at risk and social policy, both at undergraduate and graduate levels. Currently, at Jackson State University she is teaching social work ethics, psychopathology, social work practice, research methods and advanced direct practice seminar. Wendy worked in a variety of settings such as community head-start, public school, drug and alcohol rehabilitation facility and department of health. Thompson has authored three peer-reviewed articles, presented at professional conferences and edited articles for the *International Journal of Environmental Research and Public Health and Drug Abuse Issue* of *The Journal of Adventist Education*. Dr. Thompson's research focus encompasses sickle cell disease, drug and alcohol abuse, domestic violence and disparities in health care.

Pomi Yun. Pomi is currently a senior at the University of Georgia pursuing a dual degree, my Masters in Public Health concentrating in Epidemiology and Bachelor's in Genetics. Pomi's interest in music has stemmed from my sixteen years of playing the piano. He is continuing to study music through a minor at University of Georgia as well. Pomi has a great interest in patient care and have no formal educational training in Music Therapy but have chosen to research in music therapy through a partnership with Dr. Roy Kennedy.

Pomi's research in music therapy is to encourage radiation centers around the United States to incorporate music during treatment sessions to decrease negative side effects and create a more relaxing environment for the patients. pomiyun@uga.edu

ISQRMM
c/o David Akombo, Ph.D.
Jackson State University Department of Music
P.O. Box 17055
Jackson, MS 39217

ISBN 978-1-5108-3189-6

CURRAN ASSOCIATES INC. proceedings.com